LIGHT

LIGHT

AN INTRODUCTORY TEXT-BOOK

by

C. G. VERNON, M.A., B.Sc.
Head of the Science Department
Bedales School

CAMBRIDGE
AT THE UNIVERSITY PRESS
1929

CAMBRIDGE
UNIVERSITY PRESS

University Printing House, Cambridge CB2 8BS, United Kingdom

Cambridge University Press is part of the University of Cambridge.

It furthers the University's mission by disseminating knowledge in the pursuit of education, learning and research at the highest international levels of excellence.

www.cambridge.org
Information on this title: www.cambridge.org/9781316619827

First published 1929
First paperback edition 2016

A catalogue record for this publication is available from the British Library

ISBN 978-1-316-61982-7 Paperback

PREFACE

THERE are probably many teachers of science who, like the author, have felt that the teaching of the elements of Light in schools is unsatisfactory. A subject which, at the university, proves to be full of lively interest, is too often at school a dreary grind of thinly disguised geometry. The author is convinced that much of value is lost in this way, and in this book has endeavoured to present an alternative approach to the subject.

Although no one would claim that the electromagnetic theory of radiation is complete, yet it offers a satisfactory explanation of nearly all the phenomena hitherto investigated, and is generally accepted. It offers, moreover, an easily understood physical basis for all the elementary phenomena encountered by the beginner. Hence it seems reasonable to use the wave concept as the natural means of approach to the study of light: and this is the basis of the present book.

The method was used by the late Professor Silvanus Thompson and the use of the term Focal Power, in the mathematical treatment of the subject, is due to him. Geometrical Optics offers an excellent method of revision of elementary work, and the alternative methods of solving simple problems, both by means of wave-fronts and rays, are given. The book covers all the requirements of the First School Examination. A chapter on Interference, Diffraction and Polarisation, without some reference to which no account of the wave-theory is intelligible, is added, together with a series of suggestions for class experiments using real beams of light, in place of the time-honoured but unconvincing "rays fixed by means of pins."

The author wishes gratefully to acknowledge assistance from various quarters. He is indebted to Messrs Macmillan and Co., Ltd., for permission to quote from Professor S. P. Thompson's *Light, Visible and Invisible* and Sir Oliver Lodge's *Modern Views on Electricity*, and for the use of Figs. 6, 19, 118 and 135; to the Society for Promoting Christian Knowledge for permission to quote from Professor J. A. Fleming's *Waves and Ripples*; to the editor of the *School Science Review* for the various extracts indicated in the text; to Messrs G. Cussons, Ltd., for the use of the illustration of the Smoke Box (Fig. 47); to Messrs John J. Griffin and Sons, Ltd., for Figs. 3, 7 and 8; to Messrs C. Baker for the illustrations of their episcope and microscope objective; and to Messrs Philip Harris and Co., Ltd., for the illustrations of the "Rugby" optical apparatus. He gratefully acknowledges the help of Mr W. J. Jones, M.Sc., A.M.I.E.E., of the Lighting Service Bureau, in the preparation of chapter thirteen, and in providing Figs. 122 and 123.

The majority of the figures not specified above were drawn by the author especially for this book; those remaining are from Sir R. T. Glazebrook's *Light*, published by the Cambridge University Press.

C. G. V.

BEDALES
October 1928

CONTENTS

PLATES

Chapter One

HISTORICAL

A STUDY of the historical development of science shows that almost every branch has passed through several well-defined phases, which in turn reflect the progress of civilisation as we know it to-day. A primitive stage, in which certain useful information was collected as a result of experience, was followed by a speculative stage associated with the Grecian philosophers. After this came a period of comparative stagnation under the sway of the schools of Alexandria and lasting until the rise of the Muslims in the seventh century A.D. Then came the experimental activity of the Arabs, lasting until the decline of their supremacy in the twelfth and thirteenth centuries; after which, for about two hundred years, there was a gradual spreading of ideas over Europe, which was emerging from the mental obscurity of the dark ages, but no marked advance in knowledge. During this time however many illustrious men were, by their writings, preparing the way for a fresh advance, notable among them being Roger Bacon the English monk. The fall of Constantinople ushered in the Revival of Learning, and in the early sixteenth century commenced that great outburst of activity which marked the beginnings of modern science. From that time until to-day worthy successors have maintained a steady advance along the lines commenced by the great pioneers.

Of all the physical sciences the study of light appears to have attracted most attention in early times. Mirrors were in use in the earliest civilisations, and the discovery of a burning-glass of rock-crystal in the ruined palace of Nimrud shows that the functions of lenses were known, at least in part. By the time of the Grecian philosophers, 500 to 50 B.C., men were familiar with the facts that light travels in straight lines, and that when reflected by a mirror the light coming away from the mirror makes with the surface an angle equal to that between the incident light and the surface. In addition they understood the use of lenses as burning and magnifying glasses; and a manuscript of doubtful origin, but attributed to Euclid, describes the properties of spherical mirrors.

The philosophers concerned themselves with speculations as to the nature of light. There were three principal schools of thought. Pythagoras and his disciples considered light to be due to a bombardment of the eye by minute corpuscles. Plato, with a touch of the mysticism that unfortunately is found associated with many of his ideas, imagined a stream of "divine fire" from the eye which, mingling with rays from the sun and corpuscles from the object looked at, returned to the eye and produced the sensation of sight. Aristotle's mighty intellect led him to suspect that light was a form of activity in the "diaphanes"—a kind of all-pervading transparent medium which has its counterpart in the ether of modern physics—and so made an approach to the ideas that are held to-day.

The principal contribution to knowledge made by the Alexandrines is contained in two manuscripts which are supposed to be the work of Ptolemy. In these the use of lenses is described, and experimental results concerning the bending of light on entering water are recorded, tables of angles showing this bending or refraction being given.

Of the Arabs, Alhazen (eleventh century) made the most important discoveries in optics. He described the structure of the eye, and gave an account of the magnifying power of lenses. This, possibly, was a step towards the invention of spectacles at a later date in Europe, where the works of Alhazen were widely read. During the following centuries knowledge of science was slowly spreading into Europe through Latin translations of Arabic works, but original discoveries were few or none. Vitello in 1270 drew up improved tables of refraction and Roger Bacon in his *Opus Majus* indicated how lenses might be used in spectacles and for a telescope, as well as for projecting enlarged pictures. It is doubtful in the extreme however if he can be called the inventor of the actual instruments.

With the Renaissance came a renewal of activity, in England, France and Italy. The telescope was invented in Holland about the year 1608. Hans Lippershey, a spectacle-maker, appears to have been the first to construct a "perspective glass" using a convex and a concave lens, but Zacharius Jansen and Adrian Metius also made such instruments. In 1609 Galileo, hearing of the new invention, worked out the theory of it and independently constructed a telescope of the same kind. He made such wonderful discoveries with it that this type of glass is always known by his name.

In 1611 Kepler designed a telescope making use of two convex lenses, and had the first so-called astronomical telescope constructed, with its greater magnifying powers. Thanks to these inventions there was a great stimulation of interest in astronomy, and epoch-making discoveries resulted.

In 1621 Willebrord Snell discovered the relationship between the angles formed by light passing from air into water, a relationship that had hitherto eluded all investigators. His work was not made known, and soon after his death Descartes published the Law in its modern form, *The sine of the angle of refraction bears a constant ratio to the sine of the angle of incidence*, as his own discovery. Opinion is divided as to whether he filched this from Snell's papers: and it is best to have an open mind on the subject. At any rate Descartes put forward a theory that light is transmitted pressure, which, while comparatively unimportant in itself, may have led Huyghens to his wave-theory.

Christian Huyghens in 1690 published a remarkable paper,

in which he attributed the behaviour of light to some kind of wave-motion. He described the principle of wave-propagation known by his name, and which is developed at length in some of the earlier chapters of this book; and assuming light to consist of waves he was able to account for reflection, refraction and several other phenomena. He could not account for colour, however, and the inability to reconcile the behaviour of waves with the formation of definite shadows led Newton, after careful consideration, to reject this theory.

Newton's most famous experiment of course is the analysis of sunlight into a coloured spectrum using a prism. He formed an idea of the composition of white light which, though it has been called in question of late, was invaluable in promoting research. His principal work was concerned with the theory of the Nature of Light. Being unable to accept the wave-theory for various reasons, he developed the Emission or corpuscular theory in a masterly manner, and gave a theoretical explanation of the rectilinear propagation of light, the formation of shadows and the laws of reflection. Refraction he explained by assuming that the denser medium attracted the corpuscles in accordance with the law of gravitation, and as a result of his analysis showed that the speed of the corpuscles must be greater in the denser medium in order to fit in with the theory. Here was a point that could be tested experimentally, but the actual test was not applied until much later. A difficulty occurred when he was faced with the problem of a surface such as glass, which partially reflects and partially refracts light. He assumed that the corpuscles passed through regular alternate phases, being repelled when in the one and attracted when in the other phase.

So great was the weight of Newton's authority that for a long time the Emission theory was supreme, and the statement is often made that his masterly exposition held up the progress of the science. This is doubtful, but certainly the wave-theory fell into obscurity. Early in the nineteenth century Thomas Young, the Foreign Secretary of the Royal Society, made some careful observations and deductions concerning the mutual action of two separate trains of waves, and discovered the principle of interference, or alternate destruction and reinforcement that occurs when two sets of waves are superimposed. He applied this to the solution of several problems such as the alternate bands of light and darkness in the fringe of the shadow of a straight edge, a fact which had been noted by Grimaldi and Newton. Young explained this and allied phenomena in terms of the interference of waves, but was ridiculed for his pains. Fresnel followed up Young's work, and in 1818 showed that all known phenomena, including the formation of shadows, could be explained assuming that light waves are transverse. He developed the complete mathematics of the subject, and designed and carried out experiments to measure the wave-lengths of light. Two striking facts were disclosed. The waves of light are exceedingly short; and difference of colour is essentially due to difference in wave-length. The table gives values of these wave-lengths for the complete range of visible light (after Silvanus Thompson).

Colour	Wave-length in millionths of an inch	Wave-length in millionths of a cm.
EXTREMEST RED	32·4	81·0
RED	26·0	65·0
ORANGE	23·3	58·3
YELLOW	22·0	55·1
GREEN	20·5	51·2
PEACOCK	19·0	47·5
BLUE	18·0	44·9
VIOLET	16·0	40·0
EXTREMEST VIOLET	14·4	36·0

It is their extreme shortness that causes light waves to

differ, at first sight, in their behaviour from other waves such as those on the sea. As will be seen later, Newton's objections were only apparent, and due to the fact that he did not realise how minute the waves actually are.

The final disproof of the corpuscular theory came when Foucault in 1850 showed that light travels more slowly in water than in air, a necessary requirement of the wave-theory and a flat contradiction of Newton.

In order to have waves there must be a medium to carry the waves. From their immense velocity, 186,000 miles a second, and their power of passing across interstellar space, they were supposed to be waves in a highly elastic solid; and so the elastic solid theory of ether, with its rather large demands on the imagination, came to the fore.

Subsequently Clerk Maxwell demonstrated that the waves of light were remarkably similar to electro-magnetic disturbances that would arise from rapidly oscillating charges of electricity. Maxwell's theory was published in a form difficult to understand and was neglected for a time. The brilliant experiments of Hertz, who succeeded in producing and examining the properties of electro-magnetic waves, with a length of several metres, confirmed Maxwell's ideas to the full.

Thus light has come to be regarded as a simultaneous electric and magnetic strain in the ether, which is propagated at enormous speed in the same way as is a transverse wave. The theory is very incomplete. But so satisfactory has the wave-theory proved in reconciling properties of light that at first appeared irreconcilable, and in showing the similarity that exists between such extremes as the curious penetrating X-rays and the enormously long waves used in wireless transmission, that we shall commence our study by investigating the behaviour of waves, and then apply what we can learn from them to help us to understand the fascinating behaviour of light.

Chapter Two

WAVES AND WAVE-MOTION

As our knowledge of the natural world has increased, men have come to realise more and more the prevalence of waves in all kinds of phenomena. When an explosion occurs, the light whereby we see it, the sound that reaches our ears and the earth tremor that accompanies it are all the result of waves in ether, air and earth respectively. Waves on water are the most familiar, and those properties which are common to all kinds of waves can be understood from a study of water waves.

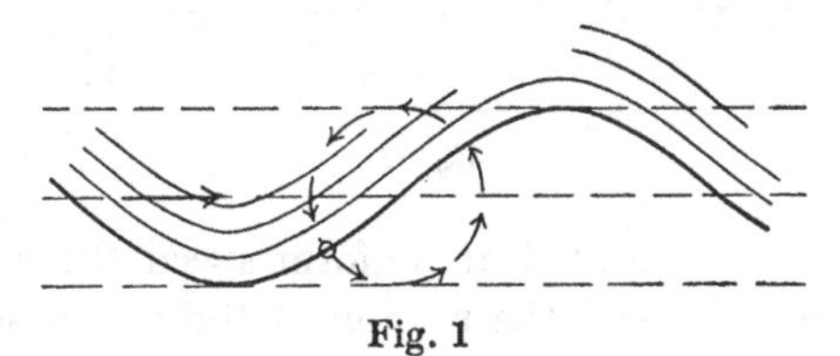

Fig. 1

Anyone who has watched the ripples that spread out on the surface of a pond into which a stone has been dropped, or who has looked over the sea from a high cliff, will have seen the apparent movement of water over the surface. But a floating object such as a cork does not move forward with the waves. At a first glance its movement appears to be merely an up and down one, although a close inspection will show that actually the object moves roughly in a circle. While passing through waves a swimmer experiences this, being carried upwards and forwards as the crest reaches him and backwards and downwards when it has passed. Now this very limited movement of the cork is similar to the movement of each of the particles of the water. The wave passes across the water, but each of the water particles in turn makes a small to and fro movement about a fixed point. The movement of the wave is quite distinct from the movements of the particle of the medium in which the wave occurs. This is best illustrated by means of models.

A thick piece of copper wire is wound into a very loose spiral and mounted with a disk of wood at one end as shown, and a small ball of plasticine is fixed on it about half-way. The wire is placed in the beam of a lantern so that its shadow is cast on a screen. As the disk is turned a wave-motion passes along the shadow of the wire, but the shadow of the ball of plasticine moves up and down in a horizontal line.

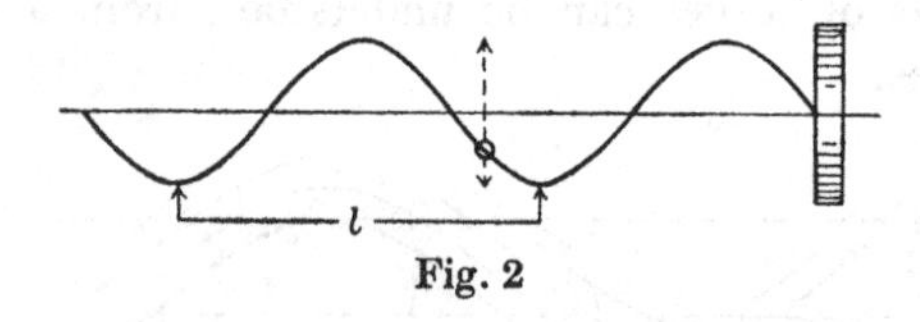

Fig. 2

When the disk is rotated at uniform speed the waves pass at a uniform rate, and the motion of the ball is seen to be similar to the movement of a pendulum. It travels backwards and forwards between two points, its speed being greatest at the middle of its path. This kind of movement, which is of universal occurrence and of the highest importance, is known as simple harmonic motion; and in general all particles of a medium through which a wave passes successively exhibit simple harmonic motion. An alternative method of showing the nature of wave-motion is described in Lewis Wright's book on Light. A grating, consisting of parallel clear spaces about $\frac{1}{16}$ of an inch apart scratched through black paint on a piece of glass, is placed in the slide carrier of a lantern. A long strip of glass is then painted black, and a wave is scratched through the paint (see Fig. 3). On passing the strip slowly through the slide carrier, while at the same time keeping the grating fixed, a wave of bright spots of light is thrown upon the screen, and moves slowly across it.

Fig. 3

In view of its importance the following demonstration of what simple harmonic motion involves is worth noting. A weight is hung from a string in the beam of a lantern. The weight is swung in a horizontal circle, and when it is spinning steadily its shadow moves with simple harmonic

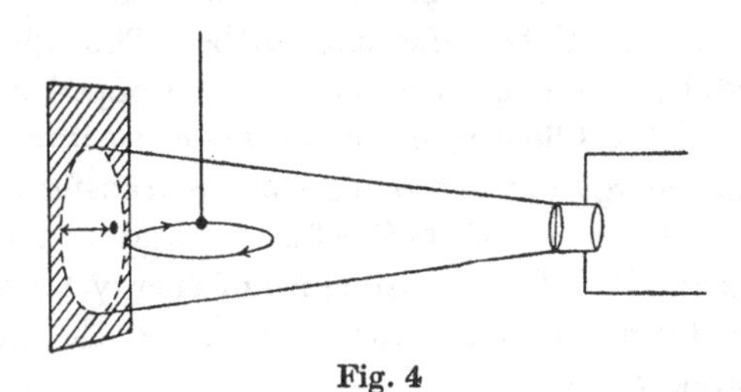

Fig. 4

motion. This helps us to understand the definition "Simple Harmonic Motion is the projection, along any diameter, of the motion of a particle moving in a circle with uniform velocity".

Returning to our shadow wave, we can now define four very important terms used in describing waves. The distance

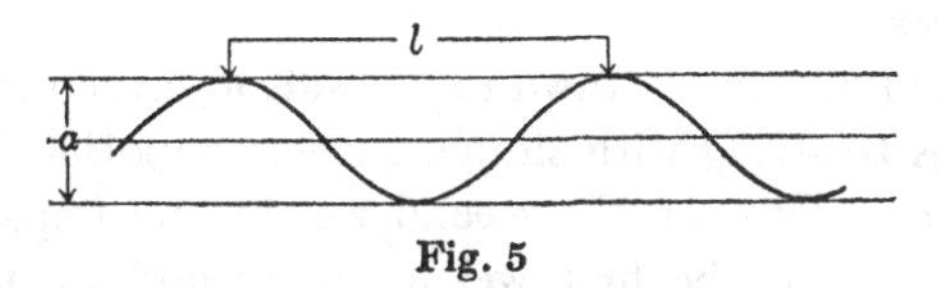

Fig. 5

between one crest and the next is known as the *wave-length.* The number of crests that pass a fixed point in a unit of time is called the *frequency* of the waves. The distance one given crest moves in a unit of time is called the *velocity of propagation* of the wave. It will be obvious that the following relations exist between these three quantities:

$$\text{Wave-length} \times \text{frequency} = \text{Velocity}.$$

For a wave of a given intensity the oscillating particles have a fixed distance between the extremes of their vibrations. This distance from crest to trough is called the *amplitude* of the wave, and the greater the amplitude the greater is the intensity.

The shadow described illustrates wave-motion. A true wave however exhibits something further. The ripples that are caused by a disturbance such as a dropped stone, on reaching any object floating on the surface cause it to oscillate. Some of the energy of the moving stone is transferred to the floating object and manifests itself as a vibratory movement of the object. It is this transference of energy to points at a distance from its source that constitutes the most important property of a wave.

A simple machine to illustrate a true wave can be constructed as follows. Strings about 2 ft. long are hung at 3-inch intervals from a rod about 8 ft. long. Alternate strings are joined in pairs, and an elastic thread is then attached to these strings. A small weight is hung on the elastic thread between each point of attachment, the complete apparatus being indicated in Fig. 6. The end weights should be heavier than the rest.

If one end weight is drawn to one side and then released it commences to swing with simple harmonic motion. But the elastic thread drags on the second weight and imparts some of the motion of the first weight to it: and by the same mechanism the second weight passes on some of its motion to the third, and so forth. Being fairly heavy each weight

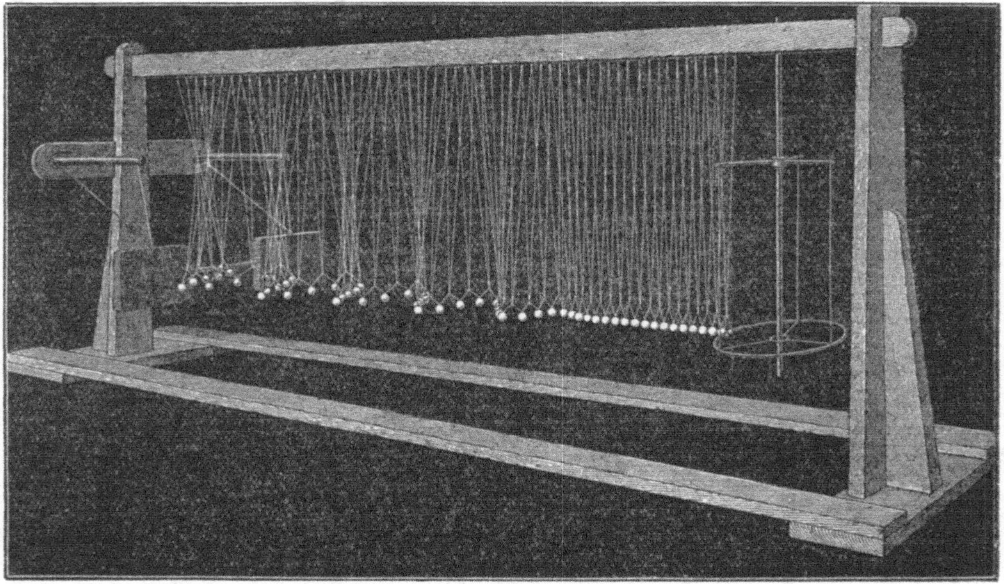

Fig. 6

continues to swing for some time after it has passed on some of its energy, and so a train of waves passes along the system and at length sets the final weight in motion.

It will be seen that this arrangement possesses two properties which are both concerned with the production of the wave, these being *elasticity* and *inertia*. If the weights were removed, *i.e.* if there were no inertia, or if a thread of soft wax possessing little or no elasticity were used for joining them, no waves would result. In Professor Fleming's graphic words "any material or medium in or on which a true self-propagating wave-motion can be made must *resist* and *persist*".* It must offer resistance to any distortion; and when it moves to return to its original shape it must "overshoot the mark or persist in movement in consequence of inertia or something equivalent to it". In both the cases described the motion of the particles is at right angles to the direction of movement of the wave. Such waves are called transverse waves, and light waves and ether waves in general are of this type. There is a form of wave in which the motion of the particles is backwards and forwards along the line of propagation of the wave, the waves then being known as longitudinal or compression waves. Sound is a result of such waves in air.

A simple apparatus for producing compression waves consists of a long helical spring suspended horizontally by threads. On giving a sharp tap to one end a pulse of compression followed by a rarefaction is seen to travel along the spring. Just as a complete transverse wave consists of crest and trough, a complete longitudinal wave consists of compression and rarefaction.

A second method of exhibiting such a wave is by means of a number of steel balls in a run-way. If a ball is allowed to run down and strike the last ball of the row, the ball at

* J. A. Fleming, *Waves and Ripples*. Published S.P.C.K.

the other end immediately flies away. The compression pulse is transmitted through the highly elastic steel at a great speed.

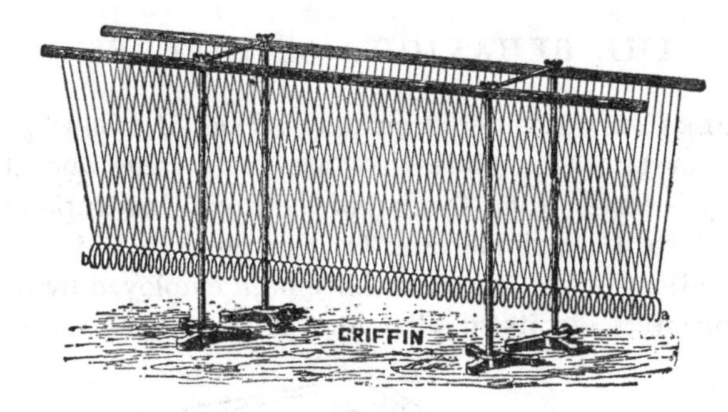

Fig. 7

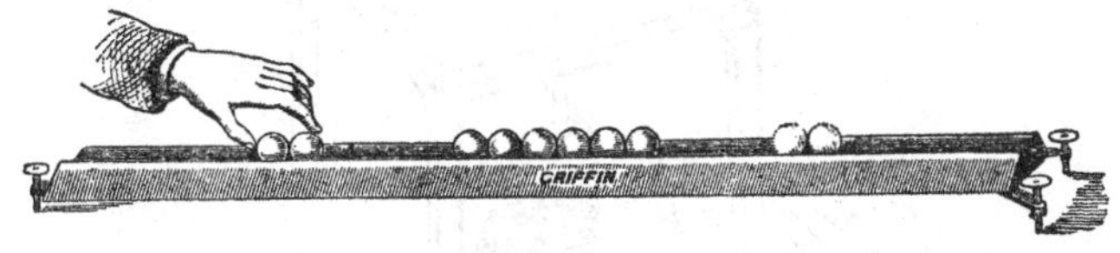

Fig. 8

It has been stated that two properties are necessary for the sustaining of waves, these being known as elasticity and inertia. In quite general terms it may be stated that the velocity of propagation of a wave is related to these, a medium with a large elasticity carrying a wave with high velocity, while a large inertia makes the velocity less.

Chapter Three

THE BEHAVIOUR OF RIPPLES

RIPPLES on water afford an easy means of following the behaviour of waves, since they are fairly large and their velocity is small. John Tyndall of the Royal Institution was the first to use a ripple tank for this purpose and various modifications of the apparatus have been employed by other experimenters. In its simplest form it consists of a shallow

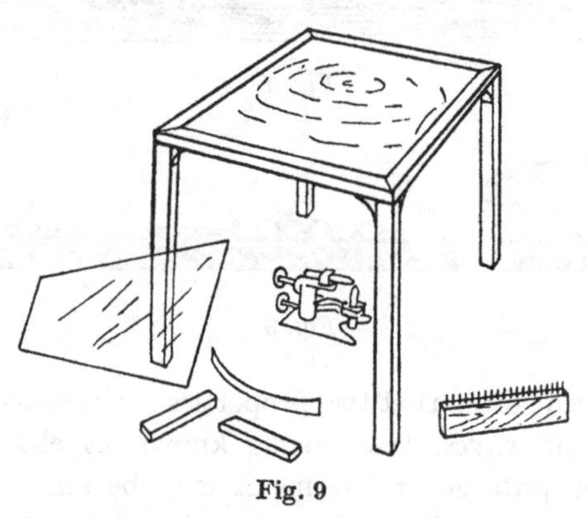

Fig. 9

wooden trough, about 2 ft. 6 in. long and 1 ft. 6 in. broad, having a glass bottom. It is supported, free from vibration, in a horizontal position and filled to a depth of a quarter of an inch with water. An arc lamp is placed underneath the centre of the tank about 2 ft. below the surface, and so a shadow of the water is thrown on the ceiling. On touching the surface of the water a train of ripples passes out from the point of disturbance and these cast definite shadows on the ceiling, the appearance being as of black lines drawn along the crests of the waves.

Using the apparatus the following can be illustrated.

(*a*) DISTURBANCE AT A POINT

If the surface of the water is touched with the tip of the finger trains of circular waves pass outwards from the point as centre, until they strike the sides, when they are reflected and a confused pattern results.

(*b*) DISTURBANCE AT A SERIES OF POINTS IN A LINE

Using a rake (see Fig. 9) which is dipped into the water, and then, when the surface is calm, withdrawn, a series of disturbances is caused at points in a line. The resultant train of ripples has a wave-front which is a straight line, parallel with the line of points.

The distance between one wave crest and the next is seen always to be constant, this distance being the wave-length. At this point it is important to realise that as the ripples are produced in two dimensions only, *i.e.* the surface of the water has length and breadth but no thickness, the waves from a point are concentric circles; but if waves are produced in three dimensions, as is true of waves of light, then from a single point the waves pass outwards as spheres, while from simultaneous disturbances from a series of points *in a plane* the wave-front produced is also a plane. Most of the problems we have to consider with respect to light concern spherical or plane waves, and these in turn can be studied using circular or straight line wave-fronts of ripples.

In both cases it will be seen that the advancing wave-front keeps its original shape, and the actual direction of the wave at any point is at right angles to the wave-front at that point. A simple experiment will suggest an explanation of these two facts.

By means of two pieces of wood a barrier can be placed across the tank leaving in the middle a gap about an inch and a half wide. Waves can be started from any point on one side of the barrier, but no matter what is the shape of the wave-front which reaches the gap a new train of waves passes out from the gap on the other side, the shape always being

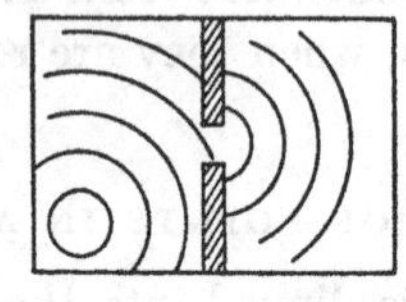
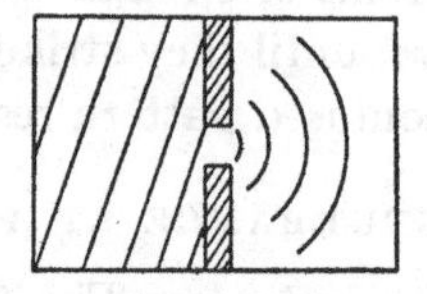

Fig. 10

circular with the gap as centre. In other words, the waves on the far side are just what would be formed by a single point disturbance at the gap. This is true moreover whatever form is taken by the waves which reach the gap, as can be shown by throwing the water into a violent agitation.

This fact is the basis of a principle that was first stated by Christian Huyghens, the Dutch astronomer, in 1678. It may

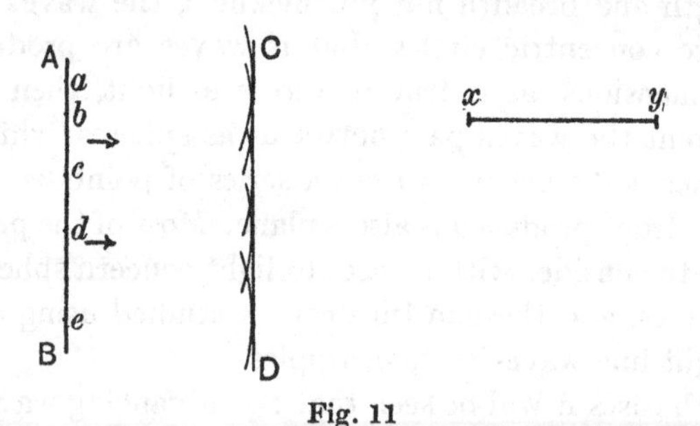

Fig. 11

be given quite simply as follows. *Each point on a wave-front acts as the centre of a new wave-train, and the resulting wave is formed by the combination of these secondary wavelets.* Let us examine what this means, using a simple geometrical construction. Let AB be part of a straight line wave-front moving in the direction shown by the arrows. Let us assume that the medium is uniform so that the wave travels with uniform velocity, and let xy represent the distance the wave moves in a small interval of time. Choose a few points at random on AB, and call them a, b, c, d, e. Now according to Huyghens' principle each of these acts as the centre of a wave, and we already know that the wave originating from a point is circular in form. Using these points as centres and xy as radius draw a series of circular arcs which represent where the secondary wavelets will have got to in the short time interval.

It will be noticed that the arcs have a common tangent in the line which we will call CD, and the more points we take in AB the more clearly does CD show up. There are of course an infinite number of points in AB, and the common tangent to all the corresponding wavelets is a straight line parallel to the original one, and at a distance xy from it. The rest of the wavelets that show in the construction would actually destroy one another by a process known as interference, which is discussed in Chapter Fourteen.

Now it will be seen that by means of Huyghens' construction we have shown that the straight line wave-front moves at right angles to its front and keeps its shape: a fact that has already been noted using the ripple tank. Imagine what would happen if a large number of men standing in a long straight rank were blindfolded and then set marching. One man alone would probably miss his direction and the possible positions he would reach after a short time could be represented by a circle about his starting-point. But if they all started together and marched at the same rate, any tendency to stray sideways would be checked by the neighbouring men, and all motion except at right angles to the rank would be *destroyed by interference.* If then the command was given to halt, and one man did not hear, once again his possible motion could be represented by a circle. In several ways the behaviour of the men resembles that of the wavelets.

Let us now consider the waves from a point. Here the similarity between our marching men and the wavelets disappears in part; for if the men started facing outwards from a point and marched at a uniform rate, it is true that their positions would always form a circle about the point, but as they marched larger and larger gaps would appear between them, whereas no gaps appear in the wave-front. Huyghens' principle helps us to understand this. Proceeding as before we can construct the successive positions of waves from a

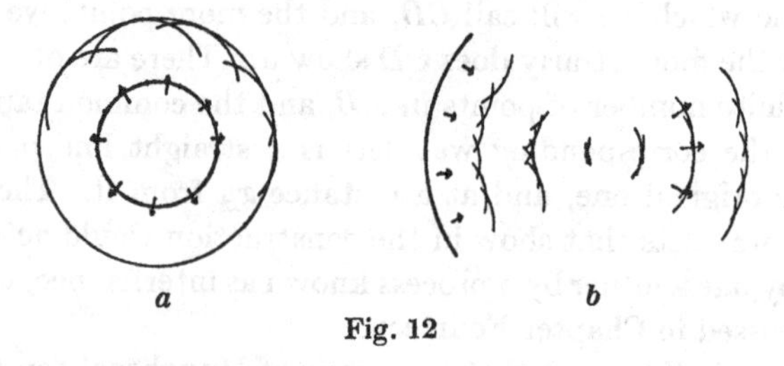

Fig. 12

point. The first position is a circle round the point with radius xy. Taking points on this circle as centres we obtain arcs whose common tangent is another circle, concentric with the first and of radius $2xy$ (see Fig. 12 *a*). This process can be continued indefinitely and as the number of points in the wave-front is infinite, no gaps will occur in the circles, which exactly parallel in behaviour the ripples on the tank.

If a wave-front is circular in form but is advancing with its hollow side forward it seems probable that it will converge on a point. By means of Huyghens' constructions this can be shown to be true (Fig. 12 *b*), and having tried this on paper it is as well to show experimentally that it does occur.

A curved piece of tinplate is placed in the tank near one end, with its hollow or concave side towards the centre of the tank. If the upper edge of the tin is tapped lightly with a pencil, hollow advancing waves are produced and these converge on a point to open out again as convex waves on the other side. Since the whole of the energy in the wave-front has to crowd through one point there will be a par-

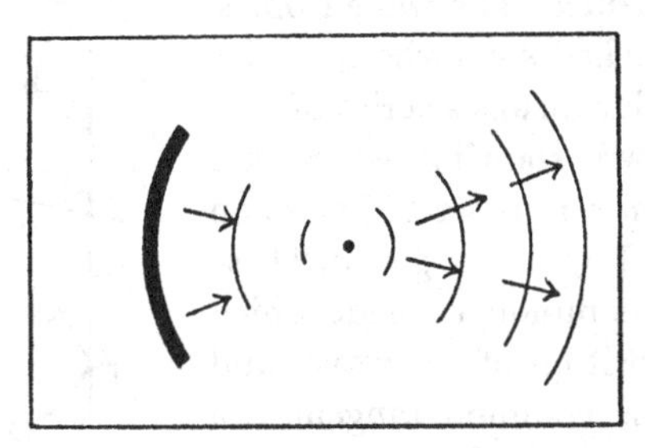

Fig. 13

ticularly violent disturbance there. Such a point is known as a focus. It may seem that a lot of trouble has been taken to demonstrate three very obvious examples, but having shown the truth of this principle in these simple cases we can employ it in the solution of more difficult problems.

REFLECTION OF CIRCULAR WAVES AT A PLANE SURFACE

Returning again to the tank, if a circular train of waves is started, when the wave reaches the side of the tank it does not stop but is reflected. By watching closely it can be seen that the reflected wave is also part of a circle, but its centre is situated at a point out in the air beyond the edge of the tank. Let us apply Huyghens' construction to this. Let P be the origin of the wave and XY the reflecting surface, and let AB and CD be two successive positions of the wave. Now the wave between E and F is drawn as though it had passed through the surface XY, and we know that this does not take place. Consider the part of the wave that has moved along the direction PQ. By the time the whole wave has reached the position $CEFD$ this part has still a distance QR to travel. *As it cannot go forward it must come back.* So with centre Q and radius QR draw an arc on the near side of the surface. The wave from Q will have reached somewhere on this arc. By similar means a series of arcs can be drawn showing where the wavelets from a number of points on the surface XY will have got to. These arcs have a common tangent which is the new position of the wave: and as is seen this common tangent is a circle whose centre lies somewhere beyond the reflecting surface. Once again the construction gives us a result in agreement with the experiment.

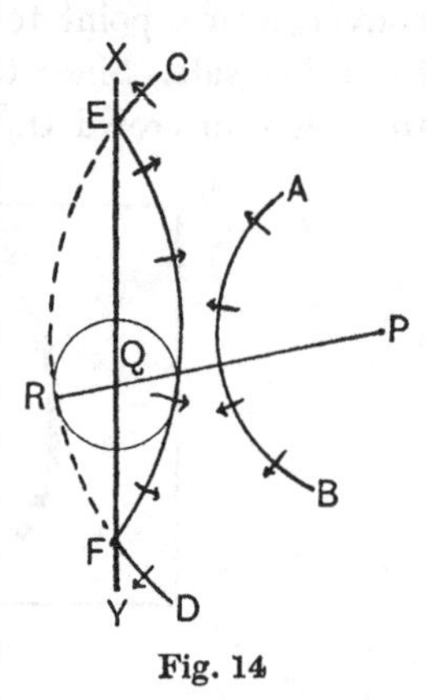

Fig. 14

REFLECTION OF A STRAIGHT LINE WAVE AT A PLANE SURFACE

A straight line wave can be produced in the tank so that it strikes one side at an angle. The reflected wave may be seen coming away from the surface, and, judging by the eye, the angle between the reflected wave-front and the surface is the same as that between the incident wave-front and the surface.

The construction is as follows. AB is the advancing wave, XY the surface, CD where the wave

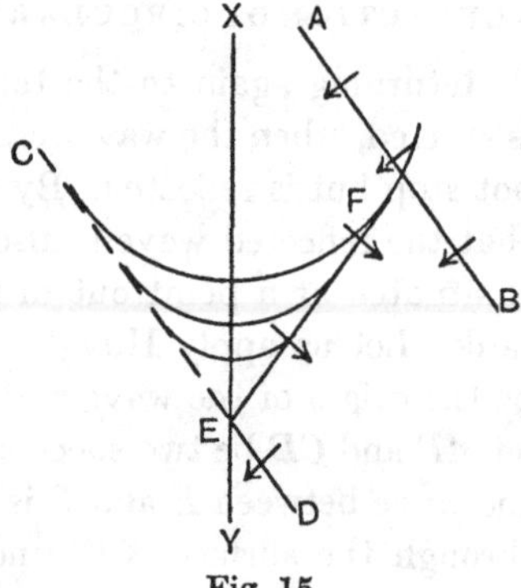

Fig. 15

would have been had XY not been there. The part CE is behind the surface and cannot exist. Arcs of circles are drawn with their centres on XY and with CE as a common tangent. The reflected wave-front is given by the other common tangent on the near side of the surface, namely EF. This is in accordance with what was seen in the tank, and it is a simple matter to show by geometry that the incident and reflected wave-fronts make equal angles with the surface. The formal proof is given in Chapter Seven.

REFRACTION OF WAVES

If the nature of the medium through which a wave is passing changes suddenly the velocity of the wave may either increase or decrease suddenly, and under these circumstances the shape of the wave also changes. Two simple cases can be illustrated in the tank. A sheet of glass shaped as a trapezium, and made to cover about half the floor of the tank, is placed in the water; and so two depths are produced and consequently two values for the speed of the ripples, which travel more slowly in the shallower water.

A circular wave started in the deeper part is flattened as it crosses the line separating the two depths, while a circular

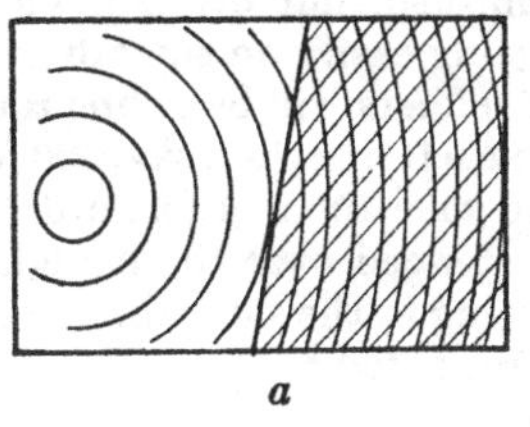

a

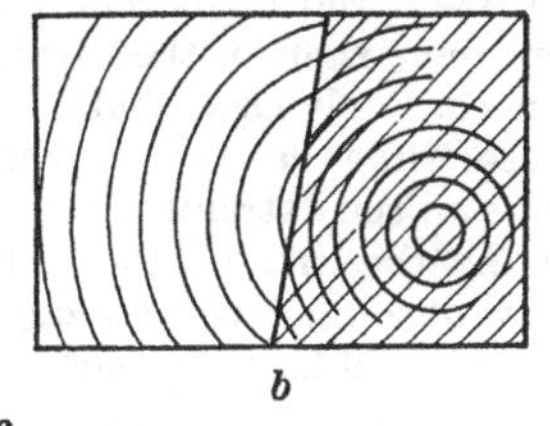

b

Fig. 16

wave from the shallower part has its curvature increased on reaching the deeper water. Thus in the former case the wave has the same shape that would result if the point of origin were farther away, while in the latter case the increased curvature is similar to that of a wave that has started from a position considerably nearer than the actual origin.

A straight line wave crosses the surface with no change other than the change of velocity if the line of separation is parallel to the wave-front. If however they are inclined there is a marked effect, to which we may give the name of refraction. If the wave starts in the deeper part of the tank the portion of the wave-front which crosses the refracting surface first is slowed down, and so the whole wave changes its direction so as to bring the wave-front more nearly parallel with the refracting surface. The greater the original angle between the wave-front and the surface, the more marked is the effect. If the wave passes in the other direction the effect is reversed, the end which passes the surface first gaining on the rest of the wave and so swinging the wave-front away from the surface. The following analogy will make this clearer.

Imagine a row of soldiers...marching over smooth grass, but going towards a very rough field, the line of separation between the smooth and the rough field being oblique to the line of soldiers. Furthermore, suppose the soldiers can march 4 miles an hour over the smooth grass but only 3 miles an hour over the rough field. Then let the man on the extreme left of the line be the first to step over the boundary. Immediately he passes into a region where his speed of marching is diminished, but his comrade on the extreme right of the row is still going easily on smooth grass. It is accordingly clear that the line of soldiers will be swung round because, while the soldier on the extreme left marches, say, 300 feet, the one on the extreme right will have gone 400 feet forward; and hence by the time all the men have stepped over the boundary, the row of soldiers will no longer be going in the same direction as before—it will have become bent, or refracted.*

Further discussion of these effects is reserved for the next chapter where the behaviour of light is dealt with.

* J. A. Fleming, *Waves and Ripples*. Published S.P.C.K.

Chapter Four

THE BEHAVIOUR OF LIGHT

As was mentioned in the first chapter, the ancients were aware that light travels in straight lines. If we take a lamp, and three screens, each with a small hole punched in it, are placed one in front of the other, it is only when the three holes are in a straight line that the light is visible. Displacement of any of the screens cuts off the light.

If the lamp is not screened at all it is visible with equal intensity from any position. This at once suggests that the wave-form given out is spherical. By means of the three screens we isolate a small portion of this spherical wave, and the first conclusion that can be drawn is that the direction in which the light is moving is along the radius of the sphere, just as our circular ripple on water moves outwards.

At this stage a difficulty may suggest itself. When the light passes through a small hole, the wave ought to spread out in a new sphere with the hole as centre. If we take a small enough hole this is true. But the waves of light are so extremely short that a hole of any appreciable size is enormous compared with them. There is not merely one point from which the secondary wavelets start, but thousands. The crowded wavelets passing through the middle portion of the hole will therefore persist as part of the original wave-form that reaches the aperture. Those starting from the edges will spread out fanwise, but this spreading can be neglected as compared with what happens to the greater part of the light.

If the holes are made successively smaller and smaller the spreading out of the wavelets becomes relatively more important, and the curious result is that if we attempt to isolate a very narrow beam of light by using extremely small holes we fail, because the few secondary wavelets spread out unchecked; and a state of affairs closely resembling that obtained with the ripples is the result.

It is convenient to imagine that we are dealing with a point source of light in considering certain problems, although of course a source of light which is so small that it has no dimensions is a practical impossibility. Nevertheless for ordinary calculations the error involved can be ignored. Let us consider how illumination due to a point source will vary with the distance from the source.

Everyone knows that the farther away a lamp is placed the less effective it becomes. A point source emits spherical waves, and the principal function of the wave is to carry the energy from the source. The actual total amount of energy emitted in one pulse from the source will be distributed evenly over the spherical wave-front, and will remain constant, since the wave cannot gain energy as it travels. For the sake of simplicity we will ignore any factors that can subtract energy from the wave.

But although the total energy remains constant, the size of the surface which carries it is growing. When the wave has

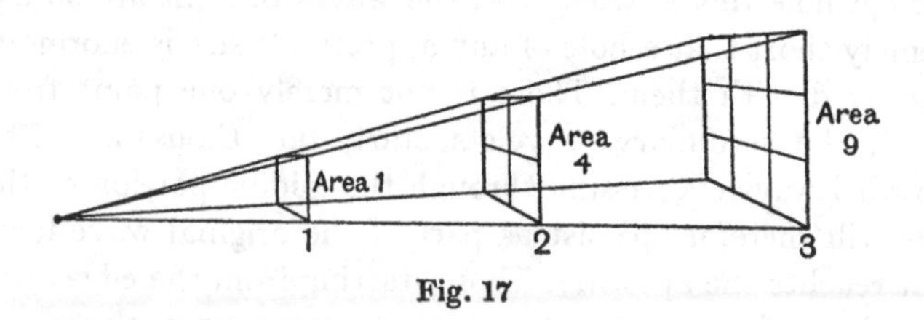

Fig. 17

travelled 1 ft. the area of the sphere is $4\pi \times 1^2$ sq. ft. or 4π sq. ft. The fraction of energy per sq. ft. is $1/4\pi$. When the wave has travelled 2 ft. its surface becomes $4\pi \times 2^2$ sq. ft. and the energy on 1 sq. ft. is then $1/16\pi$, or a quarter of what it was before. Similarly at a distance of 3 ft. the energy per sq. ft. is 1/9 of its value at 1 ft. In fact we are dealing with a very general law of physics, namely, that *the intensity varies inversely with the square of the distance.*

Let us see what this means. If a page of a book is suitably illuminated by placing a candle 1 ft. from it, how many candles must be placed 2 ft. away to produce the same illumination? Obviously 4; and at 3 ft., 9 candles would be necessary, and so on. A neat demonstration of this law of Inverse Squares for light is given in Professor Silvanus Thompson's *Light Visible and Invisible.*

The instrument for comparing intensity, known as a photometer, consists of two sheets of matt white cardboard inclined

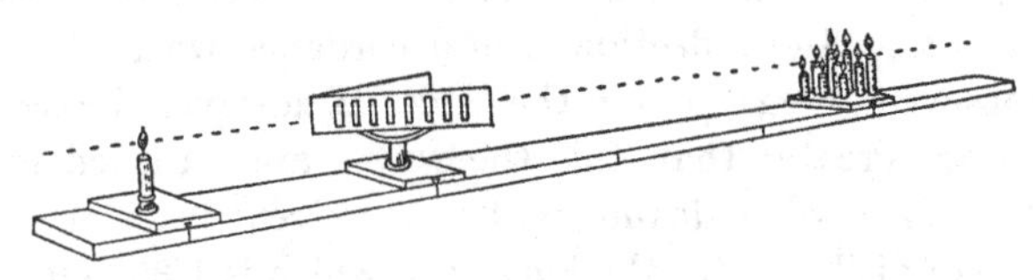

Fig. 18

at about 40°. A series of rectangular slits is cut in the front card so that the back card can be viewed through it. A single candle is placed 2 ft. away from the apex of the angle along the bisector: and 9 candles are placed 6 ft. away along this line on the remote side of the instrument. In this way the front sheet is illuminated by the single candle at unit distance, and the back sheet by the 9 candles at a distance of three units. The back of the perforated card should be blackened so as to prevent reflection from it. The middle slit should be almost invisible, those to the left appearing dark and those to the right as bright areas. Various forms of photometer have been designed, and are described in a later chapter.

By means of suitable reflectors it is possible to concentrate all the light emitted from a source into a parallel beam. The wave-front is consequently plane, and therefore remains almost constant in size, no matter how far it has travelled. Under these circumstances the intensity does not diminish; and although a certain amount of light is lost by diffuse reflection by particles of dust or moisture in the air, the beam remains almost constant and is visible at great distances. Using this device the light from a lighthouse is rendered visible up to 40 miles.

When light falls on any surface a certain amount is reflected. A black surface reflects least and polished speculum most. Reflection may be either diffuse or regular. When waves fall on a reflecting surface, provided the irregularities projecting from the surface are less than a quarter of a wavelength in size, the reflection is in accordance with what has been observed for ripples in the tank. If however the irregularities are greater than this the waves are returned in all directions; and when dealing with such extremely short waves as those of visible light, the longest of which is 1/39,000 of an inch long, this condition is the rule rather than the exception. It is by virtue of this diffuse reflection that objects are rendered visible. Regular reflection, which throws off the light in one direction only, does not render the reflecting surface visible. A striking illustration of this is afforded by painting a design in white paint on a mirror, and holding the mirror in a lantern beam. The design then appears in white against the black mirror, but where the light is reflected on to a screen a black design on a white background is seen.

When a parallel beam of light is reflected regularly from a mirror the reflected and incident beams lie in the same plane, and make equal angles with the normal to the mirror at the point of incidence. The wave-front of a parallel beam may be considered as plane, and the mechanism of reflection is similar to that already described for straight line ripples on water.

Spherical waves from a point source are reflected as

spherical waves, but after reflection they diverge from a new centre situated behind the mirror. The eye judges the origin of a spherical wave by means of the curvature of the wave which in turn is related to the distance from the centre of the sphere. Hence an eye looking towards the mirror sees an image of the source apparently behind the mirror.

Professor Silvanus Thompson's "candle and dagger" experiment is a pleasing demonstration of the position of this apparent source, which is known as the virtual image.

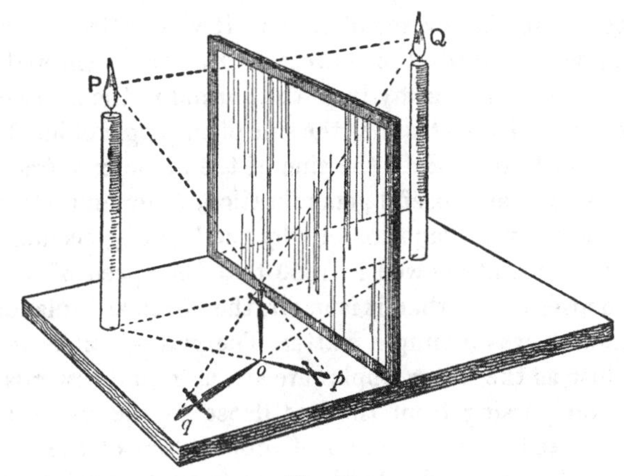

Fig. 19

A lighted candle is placed in front of a small mirror and a short dagger is stood up in such a manner that two distinct shadows can be seen, one cast by the candle and one by the mirror image of the candle. A line is drawn through the base of the candle normal to the mirror, and a second similar candle is lit and placed on this line, the same distance behind the mirror that the original candle is in front. The mirror is then removed and no change is observed in the shadows. Hence the second candle stands where the image appeared to be, viz., along the normal from the original candle to the mirror and an equal distance behind. This is a necessary consequence of the regular reflection of spherical waves, as is shown in Chapter Seven where the subject is treated mathematically. More accurate methods for illustrating the truth of these laws of reflection are given at the end of the book. The complete image in the mirror can be considered as built up of many point images, each formed in the manner described.

The third important fact that has been observed since very early times is the bending of light on passing from air to glass or water, or in the reverse direction. It was noticed that the bending was towards the normal as the beam entered the denser medium, and away from the normal in the other case. Snell's law, which states that the sine of the angle of incidence bears a constant ratio to the sine of the angle of refraction, while it serves as a useful generalisation, unfortunately does not suggest any explanation of why the bending occurs.

When Foucault showed in 1850 that the speed of light is less in denser media than it is in air, the physical explanation of refraction was a simple matter. The wave-front is swung round just as the water ripples are swung round towards the surface on passing from the less dense to the more dense medium. Hence the direction of propagation of the waves, which is at right angles to the wave-front, is bent towards the normal to the surface, exactly as the direction of the ripples was altered in the tank.

A very striking mechanical illustration of refraction can be obtained using a brass axle, 4 in. long and half an inch in diameter, and fitted with two boxwood wheels, about 2 in. in diameter, capable of turning independently. A smooth drawing-board is half covered with thick pile plush, and is tilted slightly, the plush being at the lower end. When the roller is placed on the board it pursues a straight course, and if the axle is parallel to the edge of the plush, on reaching the edge the speed is checked but the direction is unchanged. If however the axle makes an angle with the edge the roller swings round so that its direction more nearly approaches the normal to the edge. By turning the board round, the

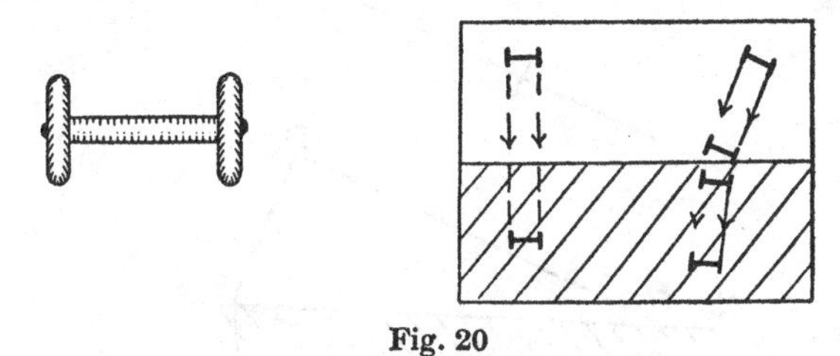

Fig. 20

change of direction away from the normal on passing from the plush to the wood is obtained. The close similarity between the behaviour of light and the roller will be obvious: but of course there is nothing in the waves corresponding to wheels and an axle.

Let us now apply Huyghens' construction to this problem, and trace the wavelets that take part in refraction. The accepted value for the velocity of light in air is roughly 186,000 miles per second, while the velocity in water is 139,500 miles per second and in crown glass 124,000 miles per second. If we take the value for air as 1, the velocity in water is $\frac{3}{4}$ and in glass $\frac{2}{3}$. This fraction is known as the *velocity constant* of the medium, and is denoted by h. Different coloured lights have slightly different velocity constants for the same medium, excepting air and a vacuum, as we shall see later.

In the diagram AB and CD represent two successive positions of a plane wave in air, and XY is the surface of a block of glass. The end C of the wave is just entering the glass, while D has still to travel a distance DE in air. Since the velocity constant for the glass is $\frac{2}{3}$, the wavelet from C will only have travelled a distance two-thirds of DE in the same time. An arc is drawn with C as centre and radius $= \frac{2}{3}DE$, and a tangent is drawn from E to this arc, meeting it in F.

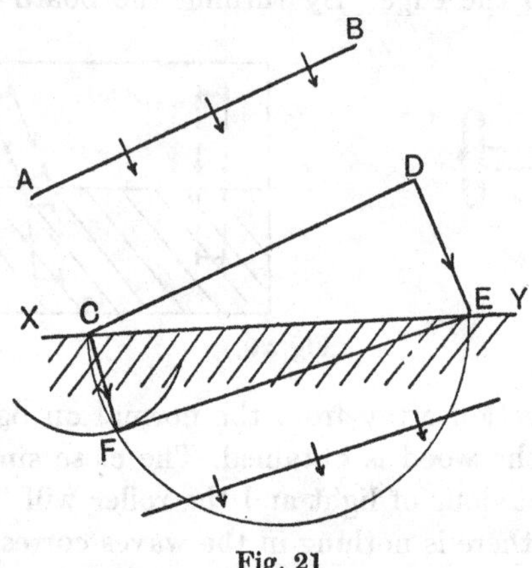

Fig. 21

Then FE is the new wave-front, which makes a smaller angle with the surface XY than did the wave CD. The construction for a plane wave passing from glass to air can be made in a similar way.

If a spherical wave starts from a point P below the surface XY of water, its successive positions may be represented by AB, CD in the diagram. Had there been no change in the medium it would later reach the position indicated by the dotted line EGF. But as each part of the wave escapes into air its velocity becomes $\frac{4}{3}$ what it was in water, and so the centre of the wave "bulges" and the curvature is increased.

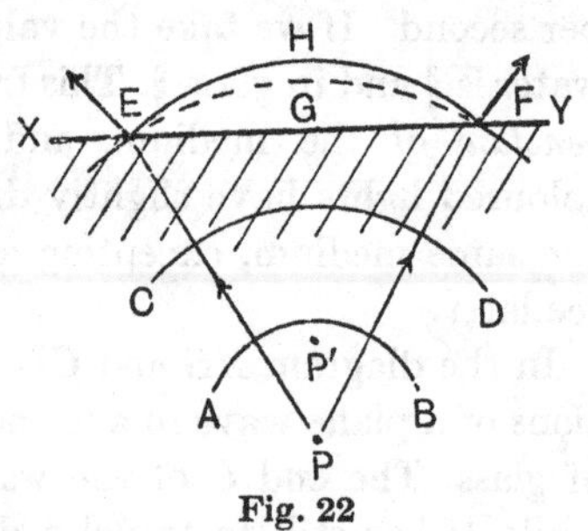

Fig. 22

The actual wave-front is shown as the line *EHF*, which is part of a circle whose centre is at P', considerably nearer the surface than the original source P. The eye judges the origin of the light from its curvature, and thus it is that a pond always appears to be much shallower than is actually the case. A more detailed account of this is given in Chapter Eleven.

It is frequently more convenient in drawing diagrams to show, not the actual wave-fronts, but the direction of movement of the waves. Lines drawn in this manner are called *rays*. They do not represent lines of light, for the narrower a beam becomes the greater is the tendency for the wavelets to spread, and it is impossible to obtain even an approximation to a geometrical line of light. Nevertheless the ray is a convenience in drawing diagrams and will sometimes be used when a wave-front diagram is unduly complex. The ray of course is always at right angles to the wave-front at that point.

When a screen containing a small circular hole is placed in front of some luminous object, such as a candle flame, an

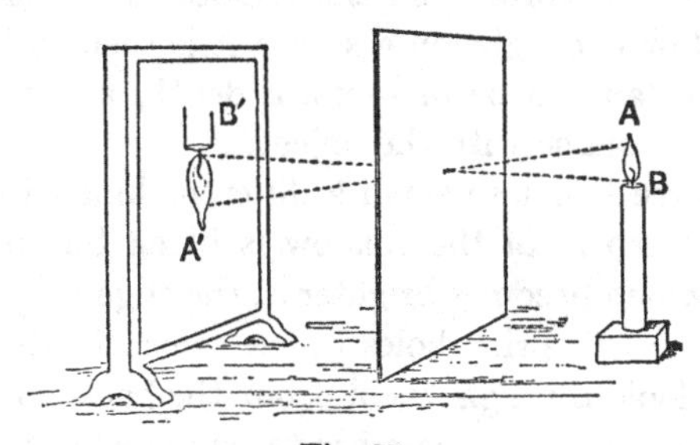

Fig. 23

inverted image can be obtained on a second screen held facing the hole. Let us see how this is caused. From *A*, the tip of the flame, spherical waves are given out, and a small portion of each gets through the hole and illuminates a tiny circular patch at *A′* on the second screen. Similarly from a point *B* near the base of the flame a small patch of illumination is obtained at *B′* *above* the other, the paths crossing at the aperture. This is indicated in the diagram. From each luminous point in the flame a corresponding illumination is formed and an inverted image is the result, the arrangement being known as the pin-hole camera. If the hole is made larger the image becomes brighter, but owing to the overlapping of the patches it is less distinct. At first sight a perfectly clear image might be expected to result from a very minute hole, but it must be remembered that under these circumstances the very narrow beams would not travel in straight lines but would spread.

If a second hole is made in the screen a second image results: and as many images are formed as there are holes in the screen. If these holes are all made into one, a blurred patch of light results from the overlapping of all the images.

When an object is placed in front of a very small source of light a shadow is formed on the remote side of the object. If this shadow is caught on a screen it is seen to have well-defined edges and to be of uniform depth, and moreover it corresponds in shape with the object.

If the source is not so small a different kind of shadow is formed. The centre of the shadow is black but there is an outer zone which becomes brighter as the edge is approached. It is instructive to prick holes in the screen in the different zones, and look through them from the back towards the source. The central zone or umbra receives no light at all, but as we move farther outwards in the penumbra more and more of the source becomes visible. This is illustrated in Fig. 24 where the rays bounding the umbra and penumbra have been indicated.

When the source of light is actually larger than the object casting the shadow, the umbra has the form of a cone with the object at its base (Fig. 25). This is an important case, as the shadows cast by the earth and moon are of this form.

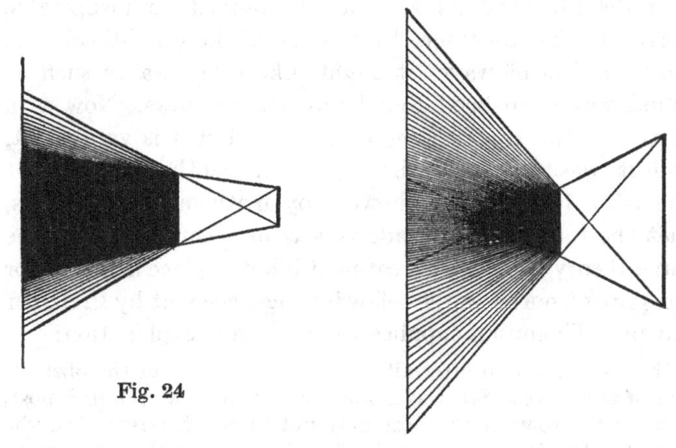

Fig. 24

Fig. 25

If the moon comes directly between the earth and the sun an eclipse of the sun is caused. If the earth is in the penumbra the eclipse will be partial, and if the three bodies at any instant are in a straight line, the rim of the sun will be visible all round the moon and an annular eclipse results.

In the much less common event of the moon's umbra passing over a part of the earth's surface a total eclipse results. The earth's orbit is an ellipse and so its distance, and consequently the moon's distance from the sun, varies. As a result the length of the moon's umbra varies; and so when total eclipses occur they may last from a few seconds to several minutes, according to the combination of varying quantities involved.

When the moon passes through the earth's umbra it is eclipsed. It should be noticed that an eclipse of the sun can occur only during the new-moon phase, while an eclipse of the moon is possible only when the moon is at the full.

The fact that light appears to travel in straight lines, and that definite shadows are formed, proved an insuperable barrier to the acceptance by Newton of the wave-theory. If light consists of waves it ought, like other waves such as sound waves, to bend round into the shadows. Now as a matter of fact this bending does occur, but it is very slight, and it was not until the early nineteenth century that Augustin Jean Fresnel showed, by mathematical analysis, that the formation of shadows was in accordance with the wave-theory. Advanced treatment is out of place here, and for our present purposes the following lucid account by Professor Silvanus Thompson supplies the necessary explanation:

...that is a question after all of the relative sizes of the obstacle and of the waves. Sea waves may meet behind a rock or a post, because the rock or the post may not be much larger than the wave-length. But if you think of a big stone breakwater—much bigger in length than the wave-length of the waves—you know that there may be quite still water behind it; in that sense it casts a shadow. So again with sound waves; ordinary objects are not infinitely bigger than the size of ordinary sound waves. The consequence is that the sound waves in passing them will spread into the space behind the obstacle. Sounds don't usually cast sharp acoustic shadows. If a band of musicians is playing in front of a house, you don't find, if you go round to the back of the house, that all sound is cut off. The sounds spread round into the space behind. But if you notice carefully you will observe that while the house does not cut off the big waves of the drum or the trombone, it does perceptibly cut off the smaller waves of the flute or the piccolo. And Lord Rayleigh has often shown...how the still smaller sound waves of excessively shrill whistles spread still less into the space behind obstacles. You get sharp shadows when the waves are very small compared with the size of the obstacle.

Perhaps you will then tell me that if this argument is correct, you ought not, even with light waves, to get sharp shadows if you use as obstacles very narrow obstacles such as needles or hairs. Well, though perhaps you never heard it, that is exactly what is found to be the case. The shadow of a needle, or a hair, when light from a single point or a single narrow slit is allowed to fall on it, is found not to be a hard black shadow. On the contrary, the edges of the shadow are found to be curiously fringed, and there is light right in the very middle of the shadow caused by the waves passing by it, spreading into the space behind and meeting there.*

This spreading, known as diffraction, is dealt with later. For ordinary purposes it is so slight that it can be neglected.

* *Light Visible and Invisible*, by S. P. Thompson, published Macmillan.

Chapter Five

REFRACTION THROUGH GLASS BLOCKS AND PRISMS

By means of an optical lantern and a slide containing a rectangular slit, it is a simple matter to obtain a parallel pencil of light, and when this falls on a screen a rectangular patch of illumination is obtained. If a cube of clear crown glass is placed in the path of the light, so that part travels over and part through the block, so long as the surface on which the light falls is normal to the pencil the illumination on the screen is still a rectangular patch. If now the block is rotated through a small angle the lower part of the illumination, caused by light that has travelled through the block, moves away to the side. By moving the screen nearer in or farther out it is easily seen that the two pencils of light are parallel, since the two illuminations remain a constant distance apart.

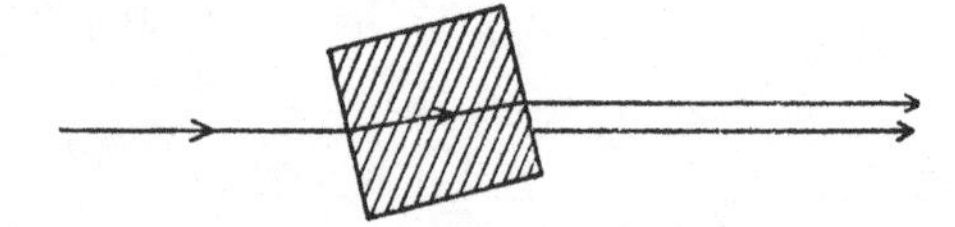

Fig. 26

A further fact may be illustrated by rotating the block through a slightly larger angle. The greater the angle between the incident light and the normal to the glass surface, the greater is the displacement. If a second block of different size is substituted, it can be shown that the displacement is also dependent on the thickness of glass through which the light passes.

The roller already described will help us to an understanding of what occurs. A parallel-sided strip of plush is fixed across the middle of the drawing-board which is then inclined as before. As the roller runs down the board, so long

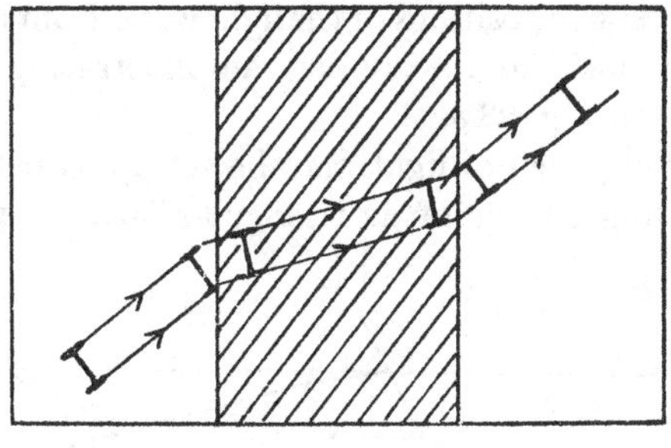

Fig. 27

as its direction is normal to the edge of the plush the only change that occurs is a slowing down on reaching the edge. If, however, the axle makes an angle with the edge, the roller is swung round on passing from the board to the cloth so as

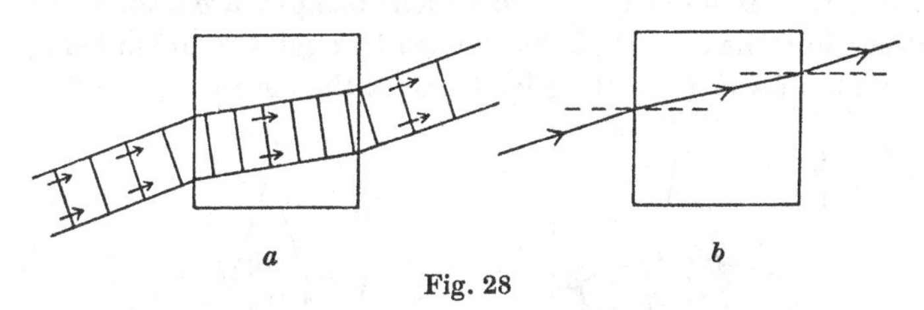

Fig. 28

to decrease the angle. But when the other edge of the plush is reached, the wheel which formerly was checked by reaching the cloth first now escapes first on to the board, and therefore begins to move faster before the other one. In this way a second change occurs, and the final direction of the roller is parallel with its original one; but there is a lateral displacement. The greater the angle between the axle and the edge of the cloth, the greater is this displacement; and by using different widths of cloth the complete agreement between the behaviour of the roller and the light can be illustrated. From these results we can build up sketches such as Fig. 28 *a* to show the successive positions of the light wave-fronts in passing through the block; and a ray diagram illustrating the direction is given in Fig. 28 *b*.

The parallel pencil of light may be set up as before and a prism, standing on one of its triangular faces, is put in the

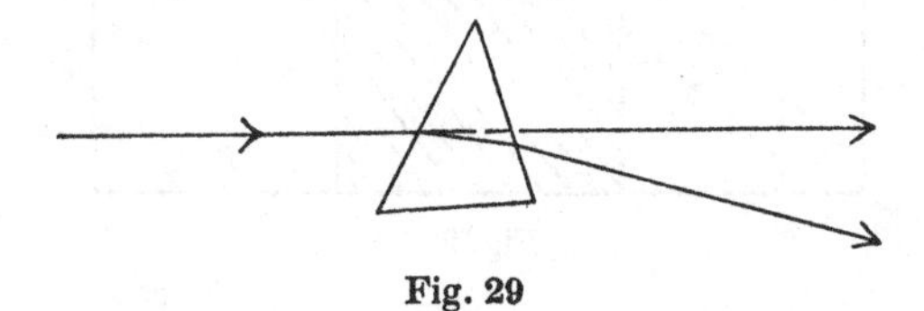

Fig. 29

place of the cube of glass. A prism with a small angle is best, and it is noticed that a very wide separation occurs between the two patches of illumination on the screen. Moreover if the screen is moved the two pencils of light are seen to be diverging, the one which has passed through the prism being deviated away from the edge towards the base.

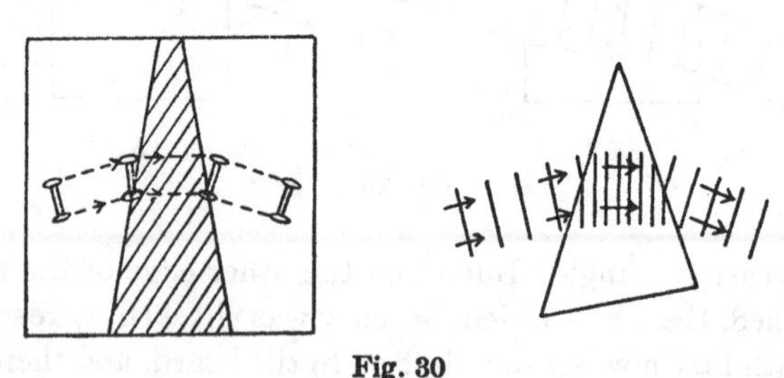

Fig. 30

When the prism is rotated slightly it is found that the deviation is varied, but for one particular position it is least, and any movement of the prism in either direction increases the deviation. This position of minimum deviation, which is obtained when the incident and emergent beams make equal angles with the respective faces of the prism, is of great importance.

Once again we can follow what is occurring by means of the roller and a triangular piece of cloth; and Fig. 30 illustrates what is seen and also the successive positions of the waves of light in passing through a prism.

In the two cases dealt with the final path of the light depends on two refractions, one occurring at each surface. If however a semicircular glass block is used, and the light is made to pass through the middle point of the straight edge, then no matter what its direction in the glass it always passes through the curved surface along a radius, and consequently does not alter its direction a second time.

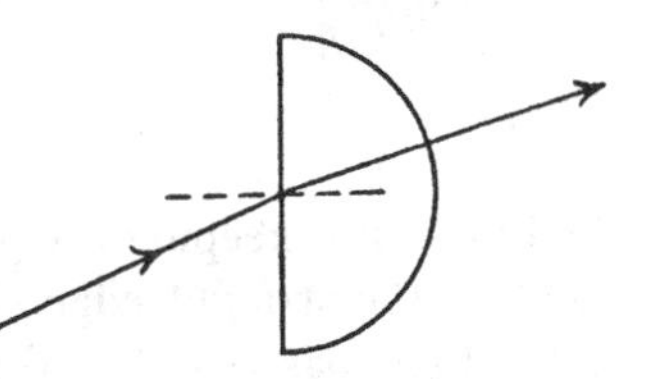

Fig. 31

Using such a block it is possible first to show the refraction when light passes from air to glass. The ray, *i.e.* the direction, changes so that it approaches nearer to the normal on passing from air to glass. Moreover if the angles between the incident ray and the normal, and the refracted ray and the normal, are measured, then the following relation is found to hold:

$$\frac{\text{The sine of the angle of incidence}}{\text{The sine of the angle of refraction}} = \text{a constant.}$$

This constant, denoted by μ and known as the *refractive index* of the medium, has a value of 1·5 for crown glass, using yellow light. For water the value is 1·33. A few other values are given overleaf. The generalisation was first stated by Snell in 1621, and is known as Snell's Law. A geometrical construction based on this for finding the direction of the refracted ray is given in Chapter Eleven.

Diamond	...	...	...	2·42
Carbon disulphide		...	...	1·63
Turpentine	...	...	...	1·46
Rock-salt	...	...	...	1·54
Flint glass	...	...	...	1·70

In order to study the refraction which occurs when the light passes from glass to air it is necessary merely to reverse

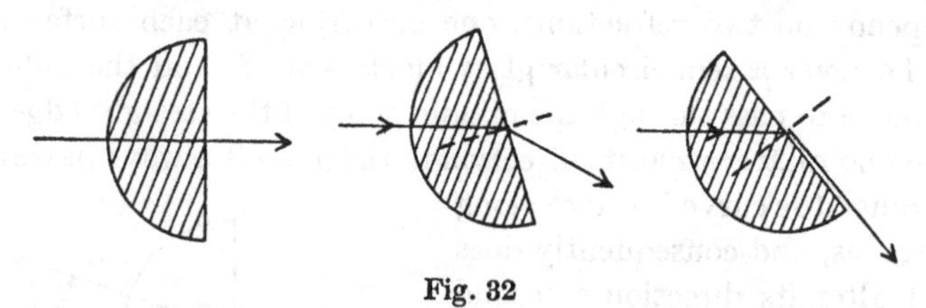

Fig. 32

the block, still keeping the path of the light through the mid-point of the straight edge. Deviation is produced then only as the light passes out of the glass. When the path is normal to the straight edge no turning effect occurs.

As the block is rotated the light is seen to be bent away from the normal, and the angle between the refracted ray (in air) and the normal is greater than that between the incident ray (in glass) and the normal. The refractive index in fact is $\frac{2}{3}$ or $1/\mu$. As the rotation is continued a point is reached when the angle in air is 90° and the light which emerges runs along the surface of the glass. This is known as the *critical position*, and the angle between the incident ray (in glass) and the normal for which this occurs is known as the *critical angle*. For crown glass it is 41° 45′. It will be found that the critical angle is such that:—

$$\text{Sine of critical angle} = 1/\mu$$

for any medium. The explanation of this is reserved for a later chapter.

A slight increase in the rotation of the block causes the emergent light in air to vanish. It is seen instead to be reflected back from the straight face of the block just as from a mirror, and examination will show that the angles between the two paths and the normal are equal. The occurrence is known as *total internal reflection.* A simple experiment with the roller will help us to understand how this is brought about.

Half of a drawing-board is covered as before with plush and it is tilted so that the covered and the bare board slopes lie side by side. The roller is placed on the plush so that in running down the slope it will pass on to the bare board.

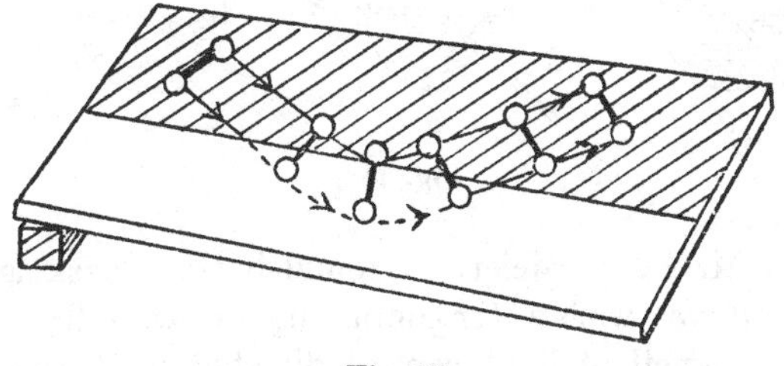

Fig. 33

If the experiment is arranged with appropriate inclination of the board and the roller, the wheel which escapes first races ahead so fast that it describes a curve with the other wheel on the inside, and actually runs back on to the cloth before the second wheel has had time to get off. The new path and the original one make equal angles with the normal. By choosing a smaller "angle of incidence" it is possible to make the roller escape from the cloth and run along parallel with the edge, thus illustrating the critical angle.

Hence we are able to obtain a rough mental picture of

what happens when light is totally reflected. The part of the wave-front which escapes first gets so far ahead that it swings right round into the denser medium before the rest of the wave gets free. The process of course is not quite so simple as this; interference plays a part in destroying secondary wavelets that are set up in air. It is interesting however to attempt to apply Huyghens' construction to the case of a plane wave passing from glass to air at an angle greater than the critical angle.

Apparatus for showing the same effects with liquids is sometimes complicated and expensive. The following method

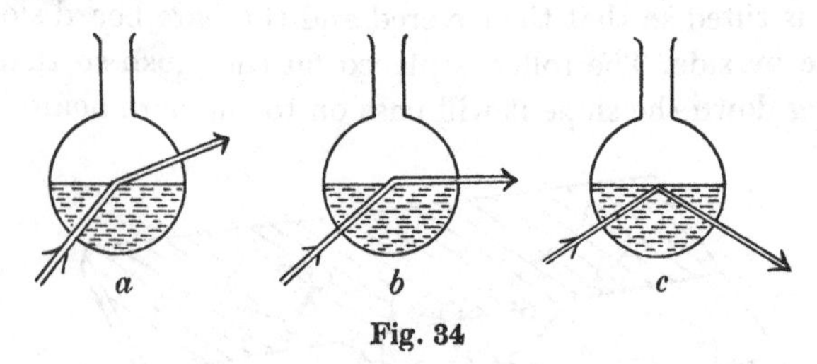

Fig. 34

is due to Mr F. A. Meier. A round-bottom flask is filled exactly half-way with water containing a trace of fluorescein. A parallel pencil of light can be directed up through the liquid to the centre of the surface, using a small mirror. By altering the positions of the flask and mirror it is easily possible to show the emergent beam in air, its extinction when total reflection occurs, and the critical position when it lies along the surface.

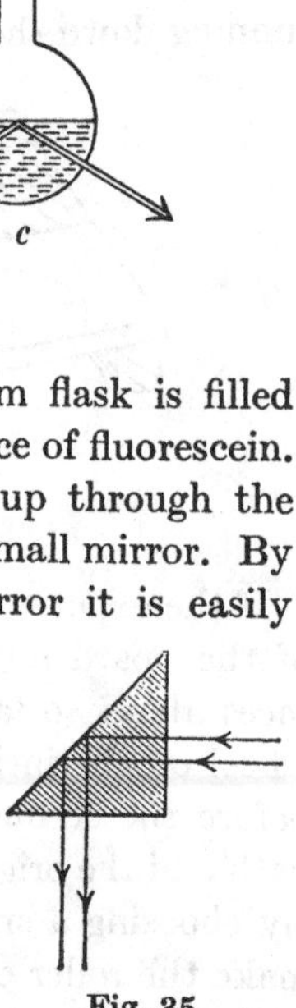

Fig. 35

An important use of total reflection is made in the construction of prismatic periscopes and compasses. If light is passed into a right-angled prism, whose other angles are each 45°, so as to be normal to one of the shorter faces, total reflection occurs at the hypotenuse, and the light emerges through the other short face, having been turned through a right angle. Such a surface possesses considerable advantages over an ordinary glass mirror. The glass in the latter is silvered at the back, and so there are two possible reflecting surfaces, namely the clear glass in front and the silver, although of course the image from the former will be very faint.

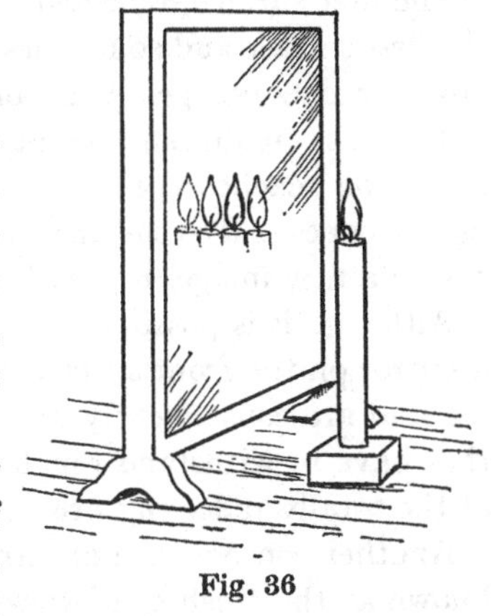

Fig. 36

But there are still further difficulties to contend with. If a candle is placed in front of a plate-glass mirror in a darkened room, and the glass is viewed obliquely, a number of images will be seen, the second being the brightest and the others gradually becoming fainter.

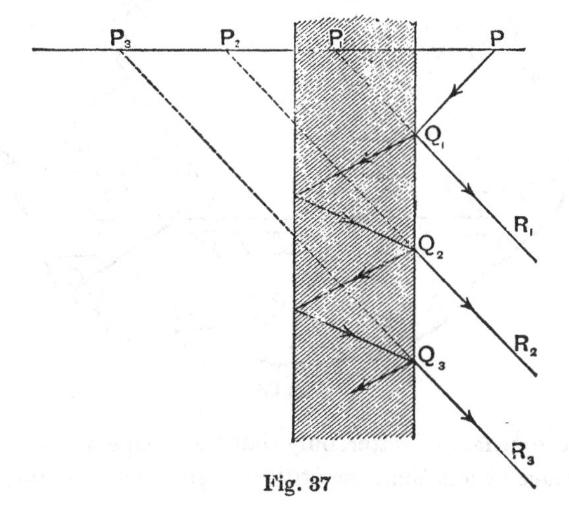

Fig. 37

The first image is caused by reflection at the front of the glass (see R_1, Fig. 37). The second, by light which is refracted at the first surface, reflected by the silver, refracted again at the first surface and so reaches the eye, R_2. Not all the light, however, does escape. A certain amount is reflected internally in the glass, again reflected at the silver and so, after another refraction reaches the air, along R_3. Yet again not all the light escapes and so the process is continued, each time giving rise to a new image, but with rapidly decreasing intensity.

Although it is possible to overcome this defect by silvering a mirror on the front surface, such mirrors have a very short life and are consequently not very practical for instruments that have to withstand rough handling. Hence the wide use of the totally reflecting prism.

Another consequence of refraction is exhibited by what is known as the "fish-eye" view. Of light given out by a point

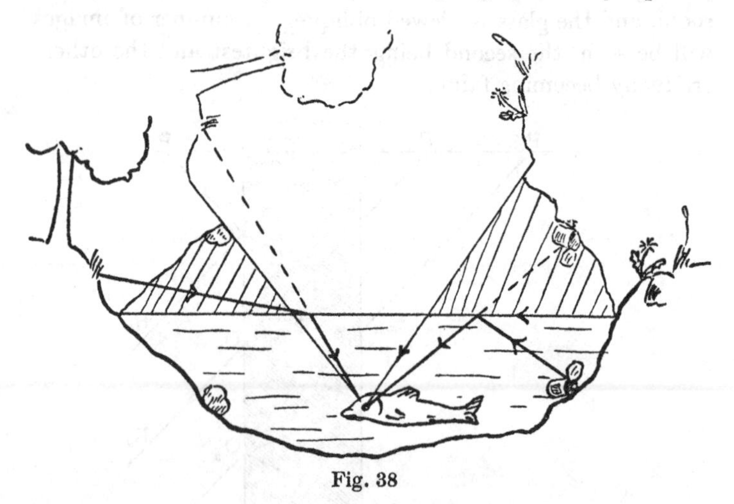

Fig. 38

below the surface of water, only that can escape which reaches the surface at less than the critical angle. The bounding lines will therefore form an inverted cone with the point as vertex, the angle of the cone being twice the critical angle. If now the path of light is reversed, all the light from above the surface of the water which reaches the point is crowded into this cone, and to an eye placed at the point the whole of the surroundings will appear crowded into a narrow circle overhead. Outside this circle the water surface will act like a mirror and reflect the bottom of the pond or container. Fig. 38 will enable this to be appreciated.

A glance at the table of refractive indices shows that for the diamond the value is very high. As a result there is a great deal of internal reflection and the stones are so cut as to enhance this. Light entering the stone can leave only in very few directions, and the intense condensation of light along these directions is responsible for the brilliance of the gem.

One other favourite example of internal reflection is the luminous cascade. A two-neck ball condenser is blackened

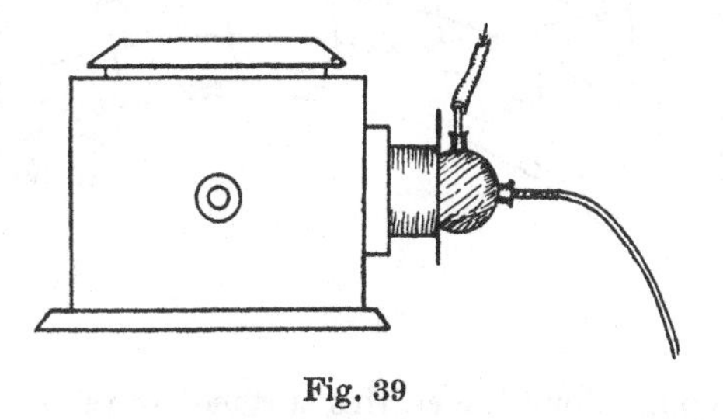

Fig. 39

except for a two-inch circle opposite one neck, and is fitted with corks each carrying a short piece of glass tubing about 1 cm. in diameter. The ball is then placed so that a brilliant beam of light enters the clear part along the axis of the jet, which is kept horizontal, the other tube being connected to the water supply. A continuous stream of water escapes in a curved jet from the nozzle, and, as the light always strikes the surface of the water at an angle greater than the critical angle, none escapes, but it is all reflected down the jet which, as a result, appears brightly luminous.

It will be noticed, by anyone who takes the trouble, that at sunset the disk of the sun appears to hover for an appreciable time just above the horizon and then swiftly to drop out of sight. Here again is an effect produced by refraction. Although, as has been stated, for most purposes the speed of light in a vacuum and in air may be taken as one and the same, namely 186,000 miles a second, there is actually a difference, and the more dense the air the slower does light travel through it. The light waves reaching us from the sun are plane, owing to the sun's distance, and when, as shown in the diagram, the sun is actually below the horizon, the

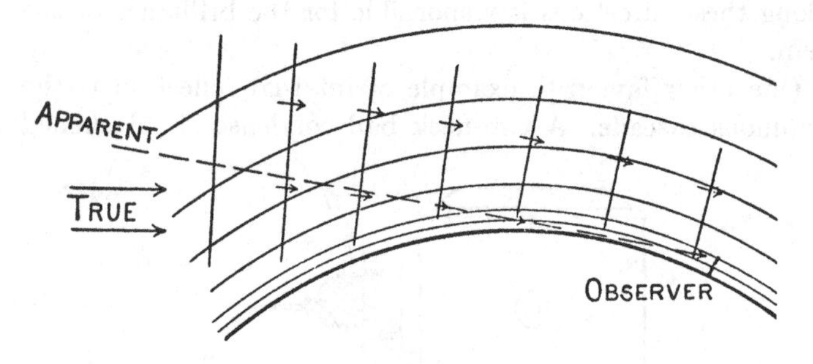

Fig. 40

waves sweeping along the earth's surface are retarded by the denser air against the surface and so swung round slightly, and appear to an observer to have come from a point above the horizon. Thus the sun is still visible for a time even when, were there no atmosphere on the earth, it would be out of sight. In a similar way if a coin is placed in a dish and the eye lowered until the coin cannot be seen, on filling the dish with water the coin again becomes visible (Fig. 41).

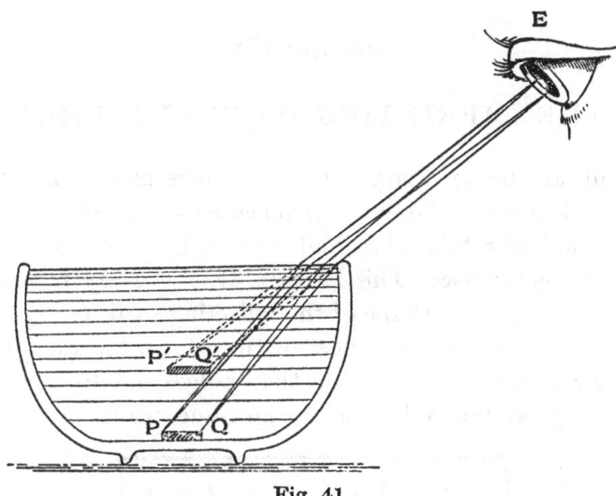

Fig. 41

A parallel phenomenon to the above is the mirage, but here it is the air close to the earth which is less dense owing to its higher temperature. Light passing obliquely downwards is hence bent up in a slight curve, and reaches the observer in such a manner that it appears to have been reflected. In this way the observer is led to believe that there is water in front of him. It is very easy to see a small mirage on the crest of a slight hill on a tarmac road when there is brilliant sunshine

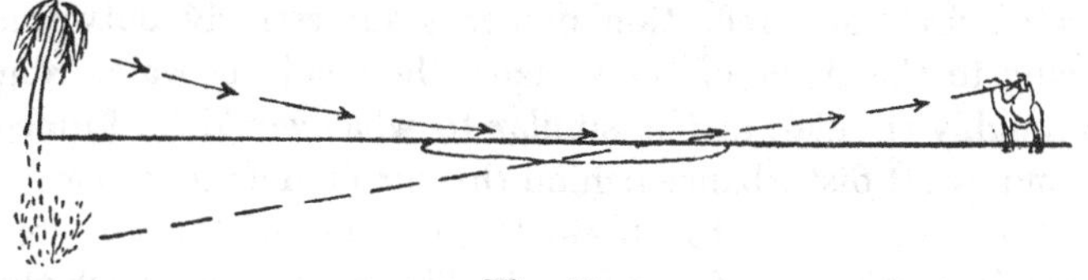

Fig. 42

on a hot day. The heat absorbed by the road raises the temperature of the air just above it; and reflections of the sky and trees so formed produce an appearance of wetness on the road.

Chapter Six

THE IMPRINTING OF CURVATURES

IT will now be necessary once again to return to the ripple tank. When a circular wave is reflected at the straight side of the tank it retains its circular form, but the direction of the wave is reversed. This latter is true of every reflection, no matter what the shape of the reflecting surface.

By means of a curved sheet of tin we can form a convex reflector in the tank, *i.e.* one that bulges out to meet the advancing waves. When a straight-line wave meets this

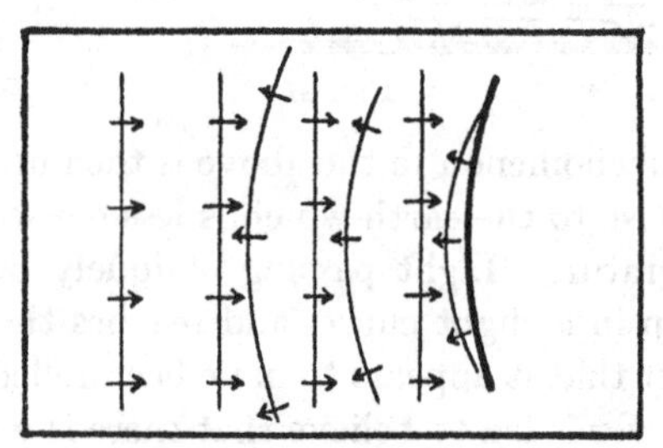

Fig. 43

obstacle it is reversed, but as the centre of the wave is reflected first, and reflection proceeds successively outwards owing to the shape of the surface, the resulting wave-front is roughly circular, and is similar to what would be formed from a point disturbance behind the mirror. The construction that has been used to obtain the position of the reflected wave is worth doing for one's self. The point from which the reflected wave *appears* to have come is known as a ***virtual focus.***

Close examination shows that the curvature of the reflected wave is apparently twice as great as the curvature of the reflector. That this is so will be shown mathematically in the next chapter. Moreover, even if the advancing wave has a curvature of its own, the mirror still imprints this curvature that is peculiar to itself, in addition to reversing the curvature originally there: as can be seen in the tank.

By turning round the tin-plate we obtain a concave reflector. This time it is the edges of the plane wave which are

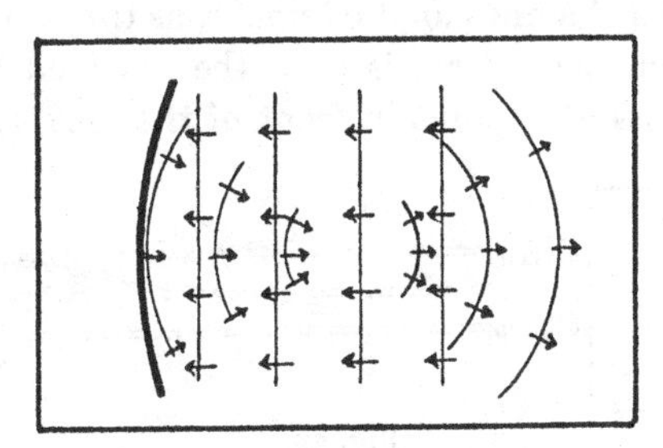

Fig. 44

reflected first, and as a result a hollow advancing wave-front is thrown back, and converges on itself to form a focus from which it spreads out again fanwise. This focus is real since the wave actually does crowd through a point. Once again the curvature imprinted is twice the curvature of the mirror, and is in addition to any curvature originally possessed by the wave.

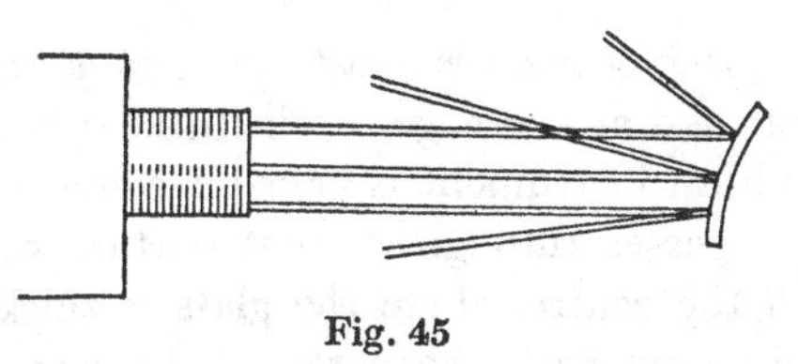

Fig. 45

It will be well now to see that light behaves in a similar way. All the objectives are removed from a lantern and then by means of suitable stops a series of parallel beams of light can be obtained. When these fall on a convex silvered mirror they are seen to be thrown back and to diverge from a point somewhere behind the mirror. A little chalk dust shaken in the air renders this visible. On the other hand a very beautiful method of showing this and many other optical phenomena has been perfected by Mr G. O. Clarke. It is known as the Smoke Box (Fig. 47) and renders visible the beams of light by means of a fine cloud of smoke, as the name suggests.

When a concave mirror is used the reflected beams converge and cross at a point in front of the mirror. The point

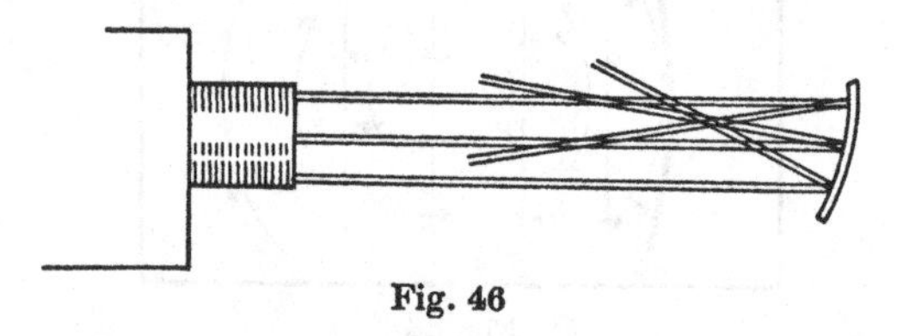

Fig. 46

to which light originally parallel is converged is known as the *principal focus* of the mirror, and its distance from the mirror is called the *focal length.*

Both these methods of imprinting curvatures depend upon reflection, and in both cases the direction of the light is reversed. When it is desired to imprint a curvature without reversing the direction it is effected by means of refraction.

A circular piece of glass is constructed so as to be thicker in the centre than at the edge, both surfaces being parts of spheres. Such an instrument is called a convex lens. When a plane wave passes through it, that portion which has to pass through the centre where the glass is thickest will be retarded, compared with the parts of the wave that pass through the thinner edges of the lens. It emerges consequently as a hollow advancing wave and passes through a focus somewhere beyond the lens. As will be seen from a

Fig. 47. Clarke's Smoke Box

diagram there is a refraction at both surfaces, since the centre of the wave is the first part to enter the glass and the last to leave.

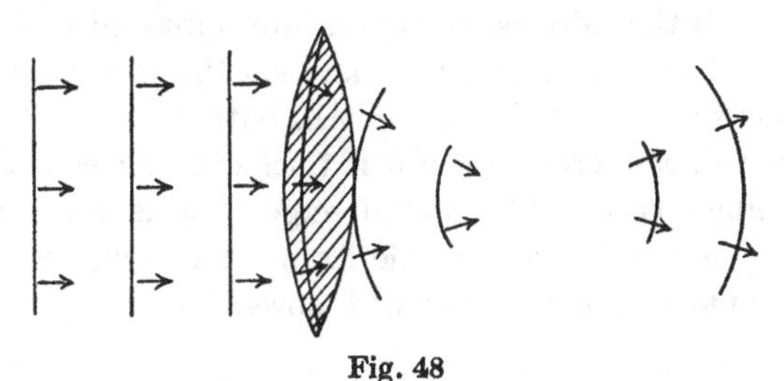

Fig. 48

If the lens is so constructed that it is thinner at the centre and thicker at the edges, *i.e.* if it is concave (Fig. 49 *a*), then

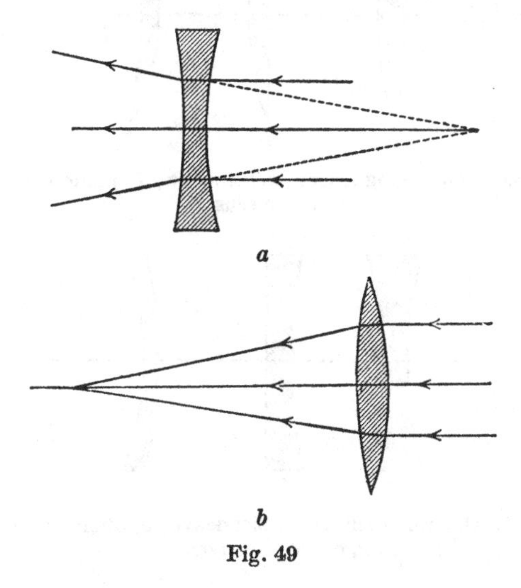

Fig. 49

it is the edges of the wave that are retarded and the light passing through is in an advancing spherical wave. That is to say, it appears to have come from some point in front of the lens, and since this is merely imaginary the point is called a virtual focus. The behaviour of both types of lenses is readily shown by one of the methods described for mirrors. Just as with the mirrors, the curvature imparted to a wave by a given lens is independent of any existing curvature. The terms focus and focal length apply in both cases.

A lens will converge light if it is thicker at the centre than at the edges (Fig. 49 *b*), and diverge if it is the reverse. Obviously a number of possible forms can exist, and these and their names are indicated as follows:

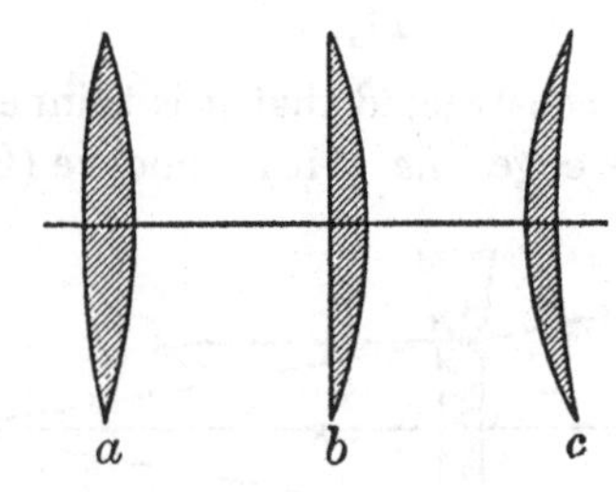

Fig. 50. Converging lenses: *a*, biconvex; *b*, plano-convex; *c*, meniscus

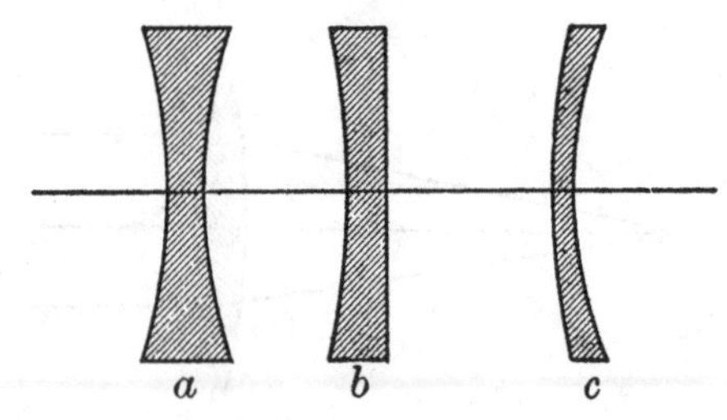

Fig. 51. Diverging lenses: *a*, biconcave; *b*, plano-concave; *c*, concavo-convex

If a convex or converging lens is mounted vertically and a luminous object is placed in front of it, an inverted real image can be obtained on a screen placed behind the lens.

Adjustment of the position of the screen makes the image sharp and clear, or "brings it into focus". On moving the lens nearer to the object the image becomes blurred but is once again made distinct by moving the screen farther away, when it will be seen that the image is larger than before. Similarly, by moving the lens away from the object and bringing the screen an appropriate distance towards the lens, a smaller image is obtained. When the lens is equidistant from the object and the screen, the image is the same size as the object.

In order to understand the formation of the image, consider first the waves from a luminous point *A* at the top of

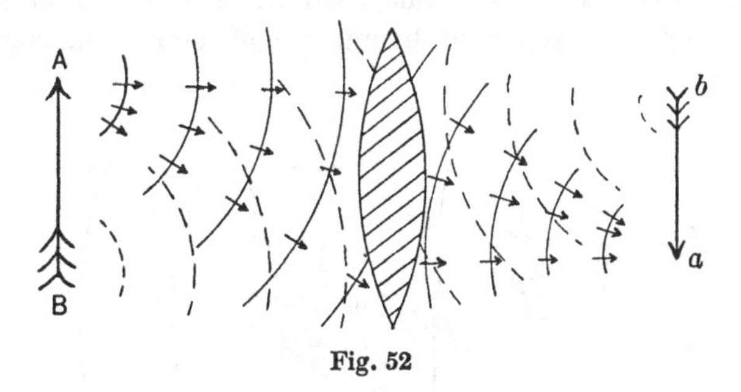

Fig. 52

the object. These will pass out as concentric spheres until they reach the lens, when the part passing through the lens will have a convergence imprinted and will converge to a focus at *a*. Similarly from *B* at the bottom of the object waves pass out to be converged by the lens at *b*, *above* the other point image. In exactly the same way every luminous point in the object has a corresponding focus beyond the lens, and if a screen is placed so as to be in the same plane as these points a sharp image is produced. By moving the screen the image is blurred, since the waves from each point in the object give rise to point images at one definite position only.

When the object is nearer to the lens the waves it emits are more diverging when they reach the lens; and so although the lens always imparts the same curvature to any wave, the resulting wave, which has a curvature equal to the sum of the original + the imprinted curvature, has to travel farther before coming to a focus. If the object is sufficiently near the lens it appears as though no image could be formed. This is in fact true and can be verified by experiment.

The use of a lens in a camera is precisely similar to the above, a real inverted image of the object to be photographed being formed on a sensitised film or plate placed at the appropriate distance behind the lens.

With a concave lens no image can be formed on a screen; but owing to its power of diverging light, and so making it

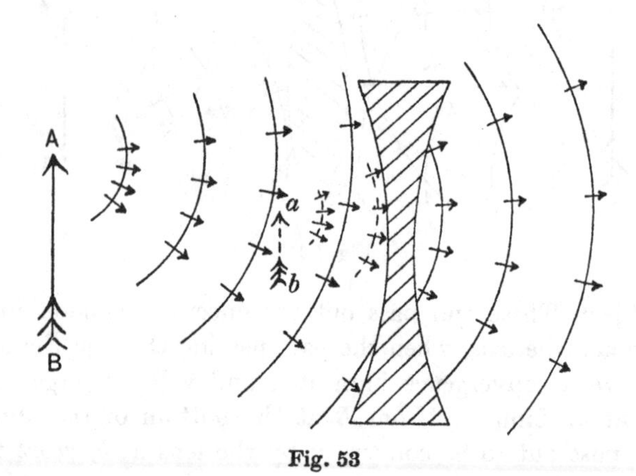

Fig. 53

appear to come from a point in front of the lens, when an object is observed through such a lens the object appears to be very much smaller than it actually is. The simplified diagram, showing what happens to waves from the top of an object, will provide sufficient explanation of this.

Chapter Seven

SIMPLE MATHEMATICAL TREATMENT

It will have been noticed in the preceding chapter that constant use has been made of the expression "curvature of the wave-front". The sole function of a lens or spherical mirror is to impart a curvature, and if one already exists the resulting curvature is the algebraic sum of the two. In order to discover what curvature can be imprinted by a lens or mirror we observe what happens to a plane wave, and the curvature imparted to such a wave is called the *focal power* of the lens or mirror.

Obviously, therefore, we need a method for reckoning curvature. Newton defined curvature as the reciprocal of the radius: that is to say, if we have three circles whose radii are in the proportion of 1 : 2 : 3, then their curvatures are in the proportion of 3 : 2 : 1. This definition, which agrees in essence with the most modern mathematical statements, is the one we shall use. It remains now to choose our unit of curvature, and this has been agreed on by an International Congress. It is the curvature whose radius is one metre and is known as the *Diopter*.

Spherical wave-fronts may be of two kinds. Either they are diverging from a luminous source or a virtual focus, or else they are converging on a real focus. In the former case we shall call the curvature negative and give it a $-$ sign, and in the latter positive with a $+$ sign.

The next thing is to make formal definitions of the different terms that are used in describing the action of optical instruments.

The *radius of curvature* of the lens or mirror is the radius of the sphere of which the mirror or lens surface forms a part, and is denoted by r.

The *principal focus* of the lens or mirror is the point at which a narrow beam, originally parallel, is made to converge, or from which it appears to diverge after falling on the instrument. The distance of this point from the lens or mirror is called the *focal length,* and is denoted by f.

Now from the definitions given above, the curvature of the lens or mirror will be $1/r$, and this is denoted by R, while the curvature imparted to plane waves, making them come to a point at a distance f from the instrument, must be $1/f$. This is called the *focal power* or *focal curvature,* and the symbol F is used for it.

In the same way u and v are used for the distance of the object and image from the instrument. When waves from the object reach the lens or mirror their curvature will be $1/u$, or U, and in order to give an image at a distance v the emergent waves have a curvature $1/v$, or V.

Whenever refraction takes place it is a result of the different speeds of light in two media. Thus one other factor that must be considered is the *velocity constant,* which is defined as the velocity of light in a medium compared with the velocity in air taken as unity. Air is the standard rather than a vacuum, since nearly all experiments are done in air. A few useful approximate values of the velocity constant, whose symbol is h, are as follows:

Using yellow light, h for water is 0·75, for crown glass 0·66 and for flint glass 0·56 to 0·61.*

For reference purposes the symbols and their meanings have been collected together in Table One.

Before proceeding to actual cases of reflection and refraction it is necessary to prove a simple geometrical theorem

* Table of values on p. 176.

TABLE ONE

Symbol	Meaning	Value
F	Focal curvature or power: positive if a convergence negative if a divergence	$1/f$ f = distance of focus from lens
R	Curvature of the surface	$1/r$ r = radius of curvature
U	Curvature of incident wave at lens, mirror, etc.	$1/u$ u = distance of object
V	Curvature of resultant wave leaving surface	$1/v$ v = distance of image
h	Velocity constant of medium $= \dfrac{\text{velocity of light in medium}}{\text{velocity of light in air}}$	$1/\mu$ μ is the refractive index

which gives an alternative method of reckoning curvature, viz. by the amount of bulge for a given length of chord.

Consider a circle of radius $OA = r$.

A chord PP' bisects at right angles a diameter AB at the point M.

It is required to show that the curvature of the arc PAP' is proportional to AM.

NOTE. AM is called the *sagitta* of the arc PAP'.

By a well-known theorem on circles,

$$AM \times MB = PM^2.$$

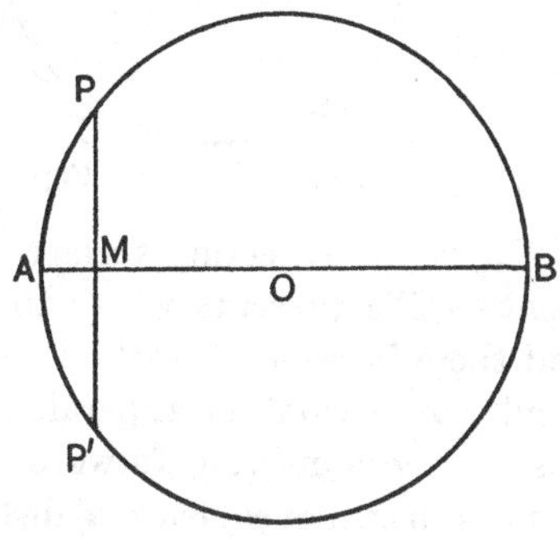

Fig. 54

If the distance PP' is very small, MB may be taken as equal to AB or $2r$ to a first approximation.

Then $$AM \times 2r = PM^2.$$

Call the distance PM unity, then

$$AM = \frac{1}{2r} \text{ or } \frac{1}{2} \times \frac{1}{r}.$$

$1/r$ has already been defined as the curvature, and so we can say, $$AM \propto \text{curvature}.$$

The constant is always $\frac{1}{2}$. Thus we get the result: *For small areas of spherical surfaces the sagitta is proportional to the curvature.* This fact is used in the construction of the optician's lens-measurer.

A. REFLECTION AT PLANE SURFACES

(i) *To show that the incident and reflected waves make equal angles with the surface.*

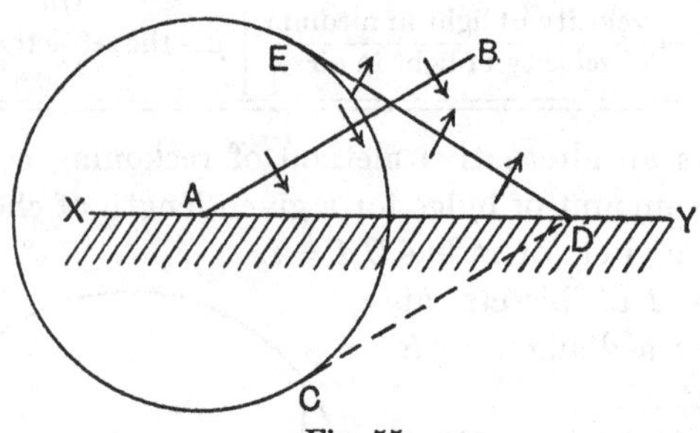

Fig. 55

XY is the reflecting surface and AB the incident wave-front. CD represents where the wave would have reached had there been no obstacle, and as the wave moves perpendicular to its front CD is parallel to AB. The part A however has not been moving forward, but the secondary wavelet from A has moved back a distance $= AC$. With centre A and radius AC draw a circle. Draw DE as a tangent to the circle. DE is the reflected wave. The angles BAD and ADC are equal, being alternate. The two tangents ED and CD

make equal angles with the line *DA* joining *D* to the centre of the circle.

Therefore $\angle BAD = \angle EDA$, *i.e.* the incident and reflected waves make equal angles with the reflecting surface. If we take the perpendiculars to the two waves and to the surface we get the expression: "The incident and reflected rays make equal angles with the normal". In this form the expression can be readily verified by experiment.

(ii) *Position of the image behind a mirror.*

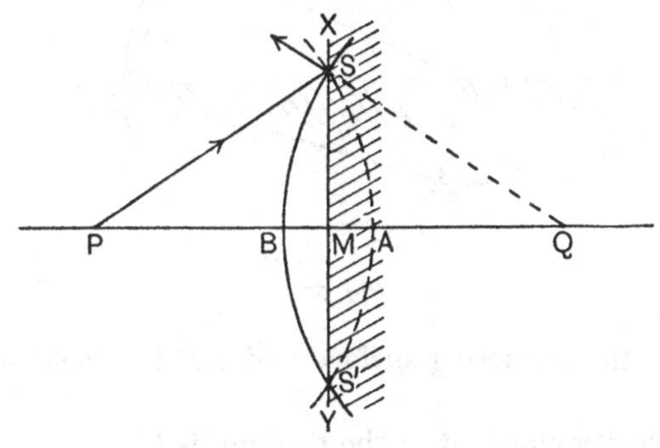

Fig. 56

XY is the mirror surface and *P* the point of origin. *SAS'* represents where the incident wave would have reached had there been no reflection. Actually the centre of the wave has moved back to *B*, such that $BM = MA$, and *SBS'* is the reflected wave whose apparent source is *Q*. Now *BM* and *MA* are the sagittae of the two curves,

$$BM = MA,$$

therefore $V = -U$ (the sign must be different since the direction is reversed)

and $u = -v.$

That is to say, the image and object are equidistant from the mirror. Moreover since *PQ* is the line of centres of the two circles, it bisects the common chord SS' at right angles. Hence we obtain the complete statement: "The image and object are equidistant from the mirror and lie on the same normal to the mirror".

B. REFRACTION AT PLANE SURFACES

(iii) *Refraction of a plane wave at a plane surface.*

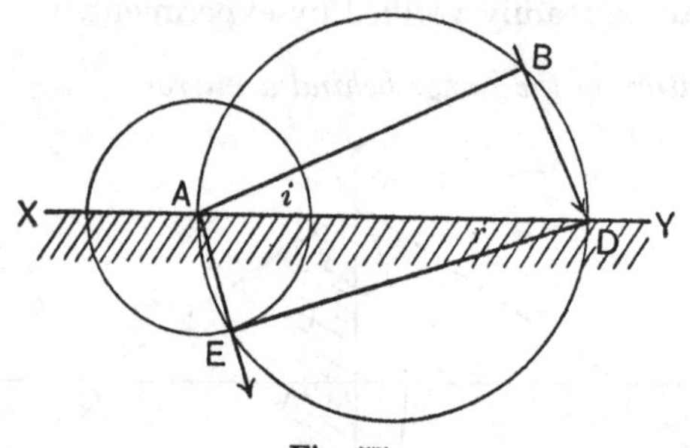

Fig. 57

XY is the refracting surface and AB the incident wave in air.

The velocity constant of the medium is h.

Draw a perpendicular through B to meet XY in D.

BD represents the direction of the incident wave.

While B moves forward a distance BD in air, A will move a distance $h \, . \, BD$ in the medium.

With centre A and radius $h \, . \, BD$, draw a circle.

Draw DE as tangent to the circle. Then DE is the refracted wave-front.

Join AE. AE is the direction of the refracted wave. Call the angle between the incident wave and the surface i, and between the refracted wave and the surface r. Then

$$\frac{\sin r}{\sin i} = \frac{AE/AD}{BD/AD} = \frac{BD \, . \, h}{BD} = h.$$

This is a constant.

The angles i and r are equal respectively to the angles between the incident and refracted rays and the normal to the surface.

Snell's Law states:

$$\frac{\sin i}{\sin r} = \mu, \text{ a constant.}$$

It will be seen that μ, the refractive index, is equal to $1/h$. Measurement of the value of μ provides an easy means for finding h.

(iv) *Refraction of a spherical wave at a plane surface.*

(*a*) Air to glass type.

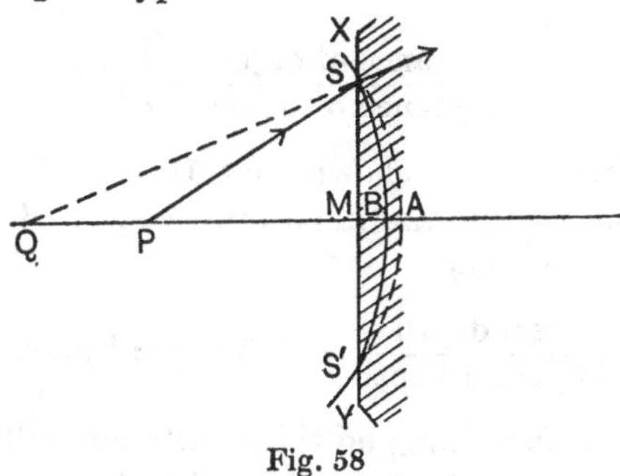

Fig. 58

The wave from P meets the refracting surface XY. SAS' represents where it would be had there been no refraction. Actually the centre of the wave is retarded and only reaches B, and so SBS' is the resultant or refracted wave, with an apparent origin at Q. Using the notation as before:

$$BM = h.AM,$$

therefore $$V = h.U$$

and $$u/v = h,$$

i.e. $$\frac{\text{Distance of object}}{\text{Distance of apparent source}} = h.$$

(*b*) Glass to air type.

If the wave originates in the denser medium, then the centre of the wave bulges on refraction and the apparent source is nearer the surface. By similar reasoning to the above

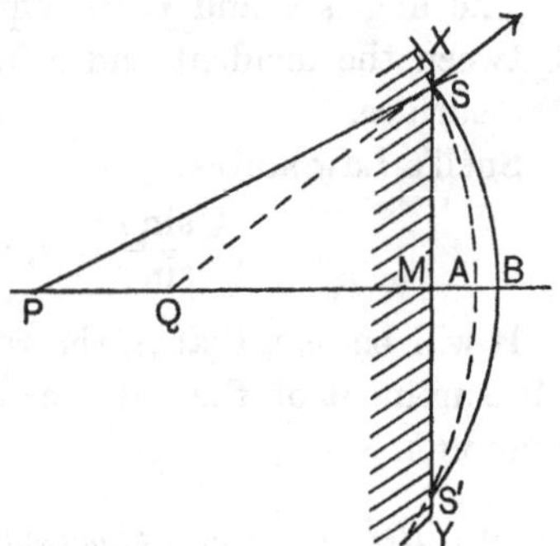

Fig. 59

$$BM = \frac{1}{h} . AM,$$

i.e. $$V = \frac{U}{h}$$

and $$\frac{u}{v} = \frac{1}{h}.$$

This time $$\frac{\text{Distance of object}}{\text{Distance of image}} = \frac{1}{h}.$$

It has already been shown that $1/h = \mu$, the refractive index. Hence we get the important case of Real and Apparent depths of a pond:

$$\frac{\text{Real depth}}{\text{Apparent depth}} = \text{Refractive Index.}$$

N.B. Proofs depending on the sagitta are only true when the angle subtended by the spherical surface is *small*. Thus the above statement for a pond is only true when viewed perpendicularly. An oblique view gives a very different result.

C. REFLECTION AT SPHERICAL SURFACES

(v) *Plane wave reflected at spherical surface.*

(*a*) Convex mirror.

SAS' is the mirror with centre O and curvature R. SMS' is the position the plane wave would have reached, but owing to reflection the centre has moved back to B, so that $BA = AM$.

From the sagittae,

$$BM = 2AM,$$

therefore $$V = 2R$$

and $$v = r/2.$$

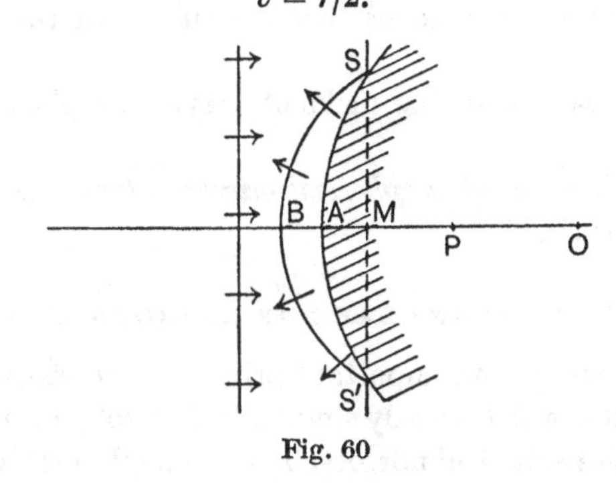

Fig. 60

That is to say, the virtual focus is situated half-way between the mirror and the centre of curvature.

(*b*) For a concave mirror the reasoning is similar.

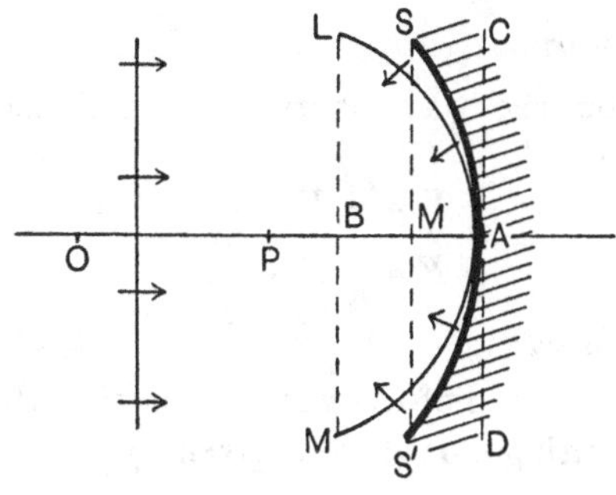

Fig. 61

O is the centre of the mirror *SAS′* as before.

CD is where the wave would have reached, *LAM* is the reflected wave.

Again $$BA = 2MA,$$

$$V = 2R$$

and $$v = r/2.$$

This time it is a real focus that is situated at the half-way point.

Since in each case the incident wave was plane, we can now state:

The focal power of a spherical mirror is twice the curvature of the mirror itself.

D. UNIVERSAL FORMULAE FOR MIRRORS AND LENSES

We now come to an important principle, the super-position of curvatures, which greatly simplifies the problems presented by lenses and spherical mirrors. This principle may be stated as follows:

The curvature of the resultant wave is the algebraic sum of the curvature of the incident wave and the focal curvature of the lens or mirror.

(vi) *For a mirror.*

On reflection the wave is reversed and then the focal curvature is added.

Hence $$V = -U + F$$

and $$F = V + U.$$

(vii) *For a lens.*

There is no reversal of the wave on passing through a lens and so the resulting curvature is given by

$$V = U + F$$

and $$F = V - U.$$

The signs of each term are important. A curve is reckoned positive if converging and negative if diverging. Hence U is generally negative, for a convex lens or concave mirror F is positive, while for convex mirror and concave lens F is negative.

(viii) *Two lenses in contact.*

If two lenses are used, the first one having a power of F_1 and the second F_2, then a wave will have each curvature imprinted on it in turn.

Hence the combined power F is given by

$$F = F_1 + F_2.$$

Each of these three formulae can now be written as reciprocals of lengths. They then read:

for mirrors $$\frac{1}{f} = \frac{1}{v} + \frac{1}{u},$$

for lenses $$\frac{1}{f} = \frac{1}{v} - \frac{1}{u},$$

for two lenses $$\frac{1}{f} = \frac{1}{f_1} + \frac{1}{f_2}.$$

Generally this will be found a more convenient form in which to use the formulae. The question of the sign must however be remembered, and this is readily obtained from the kind of curvature with which we are dealing. A few worked examples are given below showing how these formulae are used.

EXAMPLE 1. *A circular disk,* 2 *cm. in diameter, is placed* 40 *cm. from a convex lens of focal power* 10. *Where, and of what size, will the image be?*

The curvature, U, of the light from the object will be $-\frac{100}{40}$, *i.e.*

$$U = -2 \cdot 5.$$

Now $$V = U + F$$

$$= -2 \cdot 5 + 10,$$

$$V = 7 \cdot 5.$$

Therefore the distance of the image will be $\frac{100}{7 \cdot 5} = 13 \cdot 3$ cm. from lens.

Moreover $$\frac{\text{Size of image}}{\text{Size of object}} = \frac{U}{V} = \frac{2 \cdot 5}{7 \cdot 5} = 0 \cdot 33.$$

Therefore

$$\text{Size of image} = 2 \times 0 \cdot 33 = 0 \cdot 66 \text{ cm. diam.}$$

Ex. 2. *Determine the positions of the images formed when an object is placed at a distance (a) of 3 ft., (b) of 1 ft. in front of a convex lens of focal length 2 ft.*

N.B. Since all the distances are given in feet, it would be a needless complication to transform them to centimetres. Use the formula in the form

$$\frac{1}{v} - \frac{1}{u} = \frac{1}{f}.$$

In each case u is negative, since light from the object is diverging.

(*a*) $$\frac{1}{v} + \frac{1}{3} = \frac{1}{2},$$

$$\frac{1}{v} = \frac{1}{2} - \frac{1}{3}$$

and $$v = 6 \text{ ft. beyond the lens.}$$

(*b*) $$\frac{1}{v} + \frac{1}{1} = \frac{1}{2},$$

$$\frac{1}{v} = -\frac{1}{2},$$

whence $$v = -2 \text{ ft.}$$

The minus sign requires some explanation. Evidently the light from the image is diverging when it passes the lens. Hence there cannot be a real image, which would require converging light from the lens; and, moreover, since the light is not reversed in direction the virtual image formed must be on the same side of the lens as the object.

Ex. 3. *A convex lens of focal length* 10 *cm. and a concave lens of focal length* 25 *cm. are held together, and an object is placed* 50 *cm. away from the combination. Where will the image be?*

F_1 is $+ 10$, while F_2 is $- 4$.

Now
$$F = F_1 + F_2$$
$$= 6.$$

The curvature U will be $- 2$.

Then
$$V = U + F$$
$$= - 2 + 6$$
$$= 4 \text{ diopters.}$$

Therefore the distance of the image is 25 cm. from the lens.

Ex. 4. *A small object is placed* 50 *cm. from a concave mirror and an inverted image is formed* 10 *cm. from the mirror. What is the curvature of the mirror?*

In this case $U = - 2$. The light coming from the mirror is converging, *but its direction has been reversed,* and hence V also is negative.

Now
$$F = V + U$$
$$= - 10 - 2$$
$$= - 12.$$

But the focal power is twice the curvature of the mirror, and hence
$$R = - 6$$

and
$$r = \frac{100}{-6} \text{ or } - 16{\cdot}67 \text{ cm.}$$

The negative sign indicates that the curvature of the mirror is of the same order as that of the diverging light which reaches it.

Ex. 5. *A convex lens of power* $+ 8$ *is placed* 5 *cm. in front of a convex mirror. An object* 25 *cm. from the lens gives rise to an image coinciding with itself. What is the focal power of the mirror?*

Since image and object coincide the reflected light must

return along its original path. Hence the converging light reaching the mirror must have the same curvature as the mirror itself, and would come to a focus at the centre of curvature of the mirror.

It is necessary first to find v.

We know that

$$V = U + F \text{ for the lens.}$$

Therefore $$V = -4 + 8$$
$$= 4,$$

i.e. light would converge on a point 25 cm. from the lens, and consequently 20 cm. from the mirror.

Hence, for the mirror, r is 20 cm., R is 5, and F is 10.

Since this type of mirror diverges the power is -10.

Chapter Eight

MORE ADVANCED MATHEMATICAL TREATMENT

THE following problems are dealt with in this chapter:

Geometrical construction to find critical angle.
Refraction at curved surfaces.
Alteration of curvature as a spherical wave moves.
Thick lenses.
Two lenses not in contact.

E. CRITICAL ANGLE

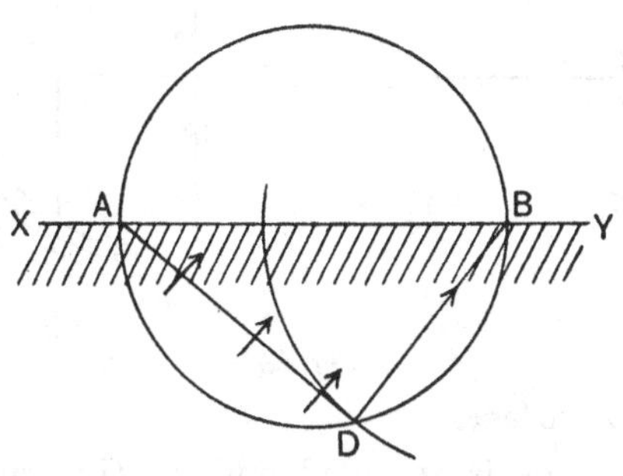

Fig. 62

XY is the surface of the refracting medium with velocity constant *h*.

Select two points *A* and *B* at any convenient distance apart, and on *AB* as diameter draw a circle.

With *B* as centre and radius $= h . AB$ draw an arc cutting the circle in *D*. Join *AD*.

Then *AD* is part of a wave-front which on emergence will pass along the surface *XY*, and the angle it makes with the surface, $\angle DAB$, is the critical angle.

A complete proof of the truth of this construction would be very complicated. It is sufficient to point out that light from A, in air, would reach B at the same instant as light from D in the medium. Any tendency to spread into the air above the surface is checked by interference.

It is important to note that

$$\text{Sine of critical angle} = BD/AB = h = 1/\mu.$$

F. REFRACTION OF PLANE WAVE AT A CURVED SURFACE

(i) *Air to glass type.*

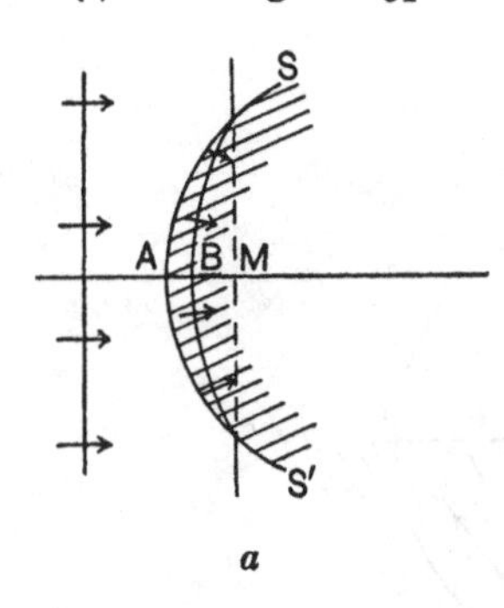

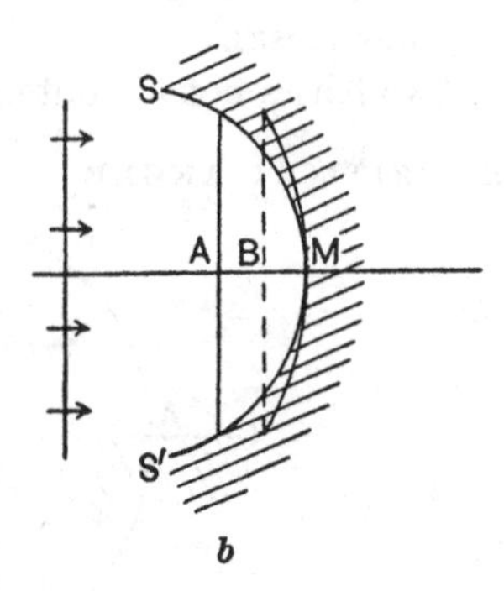

Fig. 63

(*a*) Convex surface.

Centre of wave is retarded and positive curvature imprinted.

Curvature of surface $R = AM$.

Focal curvature $F = BM$.

Now centre travels AB while edge travels AM.

Hence $$\frac{AB}{AM} = h,$$

i.e. $$\frac{AM - BM}{AM} = h$$

and $$BM = AM(1 - h),$$

i.e. $$F = R(1 - h).$$

(*b*) Concave surface—negative curvature imprinted.

Centre travels AM while edge travels AB.

Whence $$-F=-R(1-h)$$

and $$F=\quad R(1-h).$$

NOTE. The shape of the surface is the same as the shape of the curvature it imprints, and therefore the sign is the same.

(ii) *Glass to air type.*

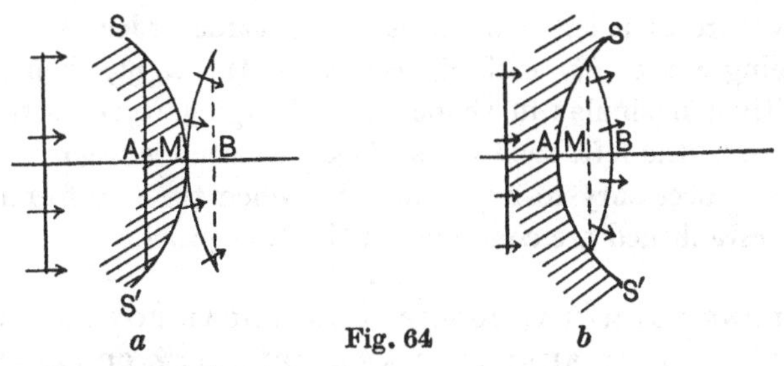

Fig. 64

(*a*) Convex surface.

The edge of the wave is advanced and a positive curvature imprinted.

As before, the curvature of the surface is R and focal power F.

Edge of wave travels AB while centre travels AM.

Then $$\frac{AB}{AM}=\frac{1}{h},$$

i.e. $$\frac{AM+BM}{AM}=\frac{1}{h}$$

and $$BM=AM\left(\frac{1}{h}-1\right).$$

Now the curvature imprinted is positive but the shape of the surface is of opposite curvature to that imprinted. Hence the sign of R must be negative.

Then $$F = -R\left(\frac{1}{h} - 1\right),$$

i.e. $$F = R\left(\frac{h-1}{h}\right).$$

(*b*) Concave surface.

The proof for the concave surface is similar. In this case the curvature imprinted is negative and therefore the surface curvature must be considered positive.

The question of sign is so important that at the risk of being redundant it is worth considering it in this way. If the curvature of a lens surface is of the same order as a converging wave, *i.e.* with its centre to the right, then it is positive; if similar in shape to a diverging wave, with its centre to the left, then the surface curvature is negative. It is very necessary to remember this when the two formulae just established are combined in the next proof.

G. LENS FORMULA, CONNECTING FOCAL POWER, CURVATURE OF SURFACES AND VELOCITY CONSTANT

Let the two surfaces have curvatures R and S. A plane wave on passing the first surface has a curvature F_1 imprinted. Any curvature on passing from glass to air is altered by $1/h$, and in addition the second surface imprints a curvature F_2.

Hence focal curvature of complete lens $= F_1/h + F_2$.

But $$F_1 = R(1-h)$$

and $$F_2 = S\,\frac{(h-1)}{h}.$$

Therefore $$F = R\frac{(1-h)}{h} + S\,\frac{(h-1)}{h},$$

i.e. $$F = (R-S).\frac{(1-h)}{h}.$$

Two very special cases which formed a discovery by Kepler in 1611 are worth noticing. Crown glass has a velocity constant of $\frac{2}{3}$. Take the case of a plano-convex lens of crown glass. Here $S = 0$ and

$$F = R\frac{(1 - \frac{2}{3})}{\frac{2}{3}}$$

$$= \frac{R}{2},$$

i.e. the focal length f is twice the radius of curvature. Similarly for a biconvex lens in which the two surfaces are equally curved, $S = -R$ and F becomes $= R$.

In its reciprocal form this very important formula becomes

$$\frac{1}{f} = (\mu - 1).\left(\frac{1}{r} - \frac{1}{s}\right).$$

H. THICK AND DIVIDED LENSES

In all the cases so far considered the thickness of the lens has been ignored. This is justifiable for most cases, and lenses which can be considered thus are known as thin lenses. The lenses used in the construction of microscope objectives, however, are certainly not thin in any sense, and so it is well to be able to establish a formula which takes the thickness of the lens into consideration. In order to do this it is first necessary to obtain a simple expression for the change in curvature of an advancing spherical wave which has moved a given distance.

(i) *The expansion of curvatures.*

Let the curvature at a point P be R ($= 1/r$). The wave travels a distance d, and so its radius becomes $r + d$.

Then the new curvature

$$R' = \frac{1}{r + d}$$

$$= \frac{1}{\frac{1}{R} + d},$$

i.e. $$R' = R\left(\frac{1}{1 + Rd}\right).$$

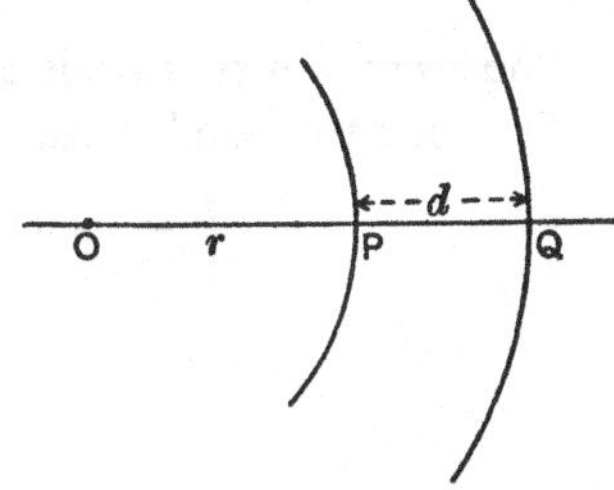

Fig. 65

If the wave is converging the curvature is increasing, and then the expression is

$$R' = R\left(\frac{1}{1 - Rd}\right).$$

(ii) *Thick lenses.*

Let the thickness of the lens be d and the curvatures of its surfaces R and S.

At the first surface a plane wave is imprinted with F_1. The wave travels the distance d and this becomes

$$F_1 . \left(\frac{1}{1 \pm F_1 d}\right).$$

At the second surface a curvature F_2 is imprinted and at the same time the existing curvature increases by $1/h$ times.

Hence $$F = \frac{F_1}{h} . \left(\frac{1}{1 \pm F_1 d}\right) + F_2$$

$$= \frac{R(1-h)}{h(1 \pm Rd.(1-h))} + S . \frac{(h-1)}{h},$$

i.e. $$F = \left(\frac{R}{1 \pm Rd.(1-h)} - S\right) . \frac{1-h}{h}.$$

(iii) *Two thin lenses separated by a distance d.*

The curvature F_1 from the first lens travels the distance d and becomes

$$F_1 . \left(\frac{1}{1 \pm F_1 d}\right) \text{ at second lens.}$$

The second curvature is added.

Thus for the combination

$$F = F_2 + F_1 . \left(\frac{1}{1 \pm F_1 d}\right).$$

Chapter Nine

THE VELOCITY OF LIGHT

UNTIL the seventeenth century it was generally supposed that the velocity of light was infinite. Galileo, it is true, attempted to measure the time that a flash from a lantern took to reach an observer a mile away, but could not detect that any time was required. In 1676 Olaf Römer, a young Dane resident in Paris, announced a surprising discovery to the French Academy. Together with Jean Picard he had carried out careful observations on the four moons of Jupiter. As each in turn passes behind the planet it is eclipsed, and, assuming that the satellites travel round their orbits at a uniform speed, the time between one eclipse and the next should be constant. This was found not to occur. At those times of year when Jupiter appears to be diminishing, this being due to the earth's movement in its orbit carrying it farther away from the planet, the eclipses occurred much later than the calculated time; while at those times when the earth was drawing nearer to Jupiter the eclipses took place too soon.

A crude mechanical example will help to show the reasoning on which Römer based his explanation of the discrepancies.

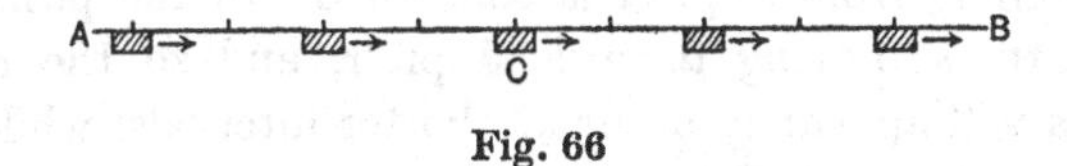

Fig. 66

Suppose there is a 10-minutes service of motor buses along a certain road marked *AB* in the diagram, and assume that the speed of the buses is 12 miles per hour. In 10 minutes a bus will travel 2 miles and so they will be spaced out at 2-mile intervals along the road. If a man stands beside the road at *C*, a bus will pass him every 10 minutes. Suppose now the man begins to walk at 4 miles an hour in the same direction as the buses, just as a bus is leaving *C*. The next bus will be 2 miles behind him, and as he is moving forward at 4 miles an hour it is *catching him up at a rate of* 8 *miles an hour*. As a result it will be 15 minutes before the bus passes the man, and so long as he continues to walk, the interval between the buses that pass him will be 15 minutes.

Suppose on the other hand the man was walking towards *A*. He and the bus 2 miles away are approaching one another at 16 miles an hour. The 2 miles will be covered in 7½ minutes,

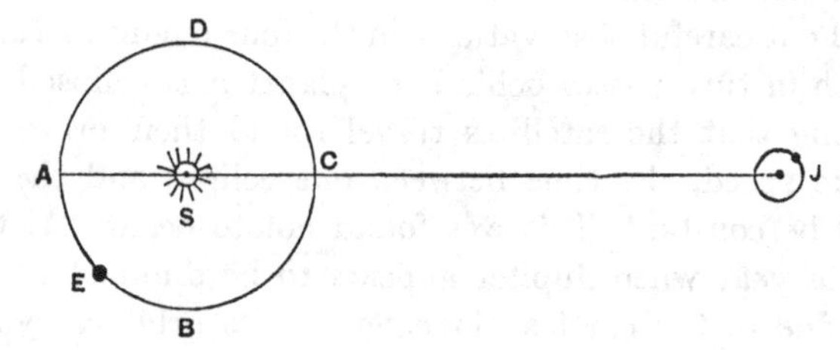

Fig. 67

and so long as he continues to walk the time interval between the buses will be only 7½ minutes. Let us now apply this to Römer's problem. When the earth is at the positions marked *A* and *C* it corresponds to the man standing still so far as light coming from Jupiter is concerned. In the position *D* the earth is moving towards Jupiter, and so the regular eclipses will appear to occur at shorter intervals, while at *B* the intervals will be longer.

Knowing the differences in the times, and the diameter of the earth's orbit, Römer calculated that light took 22 minutes to cross the earth's orbit. This is actually much too long: and as his figures were not very convincing his theory that light has a definite velocity was largely ignored.

A confirmation of his ideas came some years later. The English astronomer Bradley found that certain stars appeared to be slightly out of their calculated direction at certain times. One day whilst yachting on the Thames he noticed that every time the boat changed its course the pennon at the mast-head pointed in a new direction, showing apparently that the wind had changed. The boatman assured him that the wind was constant. The direction of the pennon in fact was determined by the relative velocity of the wind and the boat, and could easily be calculated by anyone familiar with the principle of mechanics dealing with the combining of velocities. Another similar effect must be familiar to all. On a day on which rain is falling vertically, as soon as one commences to walk in any direction the rain appears to slant down so as to beat in one's face, and the faster the walker's movement, the more oblique does the rain become. Bradley applied the idea to his astronomical problem and calculated the speed of light from the known velocity of the earth. He found it to be about 185,000 miles a second.

Both these methods have the disadvantage of depending on the accepted value of the earth's orbit, the accuracy of which may be questioned. Methods not involving astronomical observations were designed to overcome this difficulty.

The first of these was carried out by Fizeau in 1849. In principle it consisted of letting light pass through a shutter, which was opened and closed many times a second, on to a distant mirror which reflected the light back to the shutter. The speed of light being very great, the returning light would pass back through the shutter before it was closed each time, and be observed. If the rate of opening and closing the shutter were then increased, a point would be reached when the returning light would get to the shutter just after it had closed and so would not pass through. The shutter effect was obtained by passing the light by the edge of a wheel bearing rectangular teeth, so that the light could pass through the intervals between the teeth. The actual apparatus was as follows: Light from a source S passed through two lenses and after reflection at a plate of clear glass G was brought to a focus at a point P. This point was situated at the principal focus of a lens and so the light passing through O was transmitted as a parallel beam. After travelling 3 or 4 miles the beam passed through another lens Q which brought it to a focus on the surface of the mirror M, whence it was reflected back along its own path and formed an image again at P which could be viewed through the glass plate by an observer.

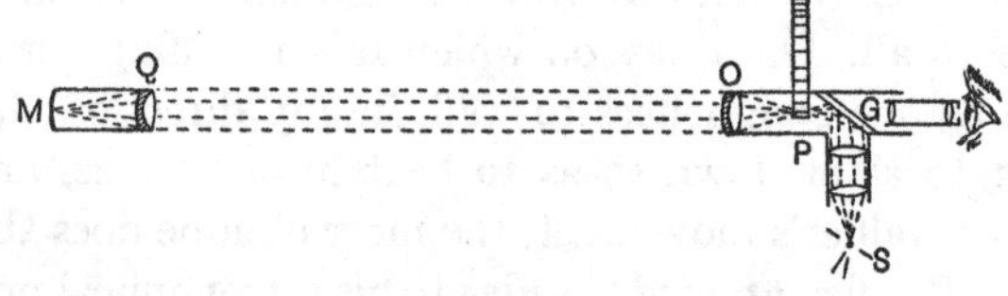

Fig. 68

The toothed wheel was so placed that as it rotated the teeth passed one after another through the point P, thus intercepting the light. When the wheel began to move a flickering image was observed, which became steady with an increase of speeed owing to the persistence of vision. A still further increase resulted in the disappearance of the image, while when the wheel moved faster still an image was again visible. In fact the light that had left through one gap returned through the next.

Measurement of the distances the light had travelled, and the dimensions and speed of revolution of the wheel, gave figures from which the velocity of light could be found. Fizeau obtained the value 315,000,000 metres a second, but Cornu, with improved apparatus, gave a value of 300,400,000 metres a second.

Soon after this the results of a different method were made

public. In 1838 Arago suggested to Fizeau and Foucault a method whereby the velocity of light could be measured both in air and in water, thus putting the emission theory to its final test. Together they commenced work, but before the details were settled they decided to work independently, and Foucault alone obtained results by this method which bears his name. Sunlight from a slit *S* was allowed to pass through a lens *L* and after reflection from a plane mirror *R* was brought to a focus at the fixed mirror *M*, 20 metres away, whence it returned along its own path. A sheet of plane glass was employed to enable the image to be observed at *O*. The

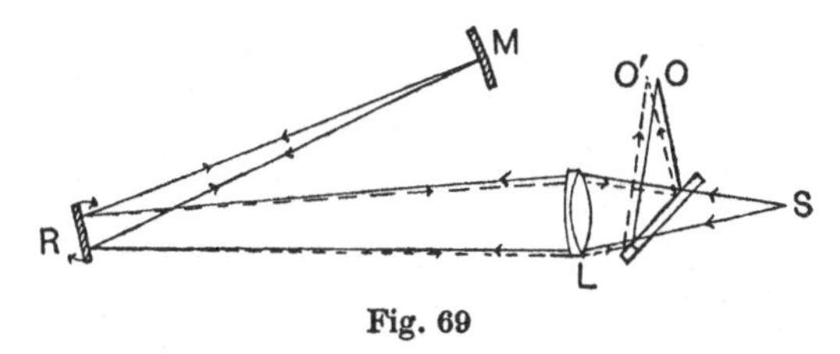

Fig. 69

mirror *R* was then rotated at high speed. A band of light was formed round the room and just at one position of *R* this fell on the mirror *M*. By the time the light had passed to *M* and back, *R* had turned through a small angle, and consequently instead of being reflected back along its previous path there was a slight shift of the image to the position *O′*. By measurement of this distance *OO′*, and of the speed of rotation of the mirror, the velocity of light could be calculated. Foucault's result was rather too low, but he performed a second experiment of immense importance. A tube of water with glass ends was placed between *R* and *M*. The shift of the image was increased considerably, and so it was shown conclusively that light travels more slowly in water than it does in air. This constituted the final disproof of the corpuscular theory.

Michelson and Newcomb both made improvements on Foucault's original apparatus, and by increasing the distance between *R* and *M* succeeded in getting a shift of several centimetres in the position of the image as compared with Foucault's deflection of 0·7 mm. Their final figure for the velocity of light was 299,860,000 metres per second. As will be seen the mean of all the experiments can be taken in round figures as 300,000,000 metres or 186,000 miles a second.

Chapter Ten

THE SPECTRUM AND COLOUR

THE name of Newton is always associated with the production of a coloured spectrum from white light by passing a narrow beam of sunlight through a glass prism. Seneca (A.D. 2–66) recorded the production of colours in this way and pointed out the similarity between the colours of the rainbow and those produced by the edges of a piece of glass. No satisfactory explanation of their production was given however, the most popular view being that red was strongly condensed and violet highly rarefied light.

In repeating Newton's experiment of throwing a spectrum on to a screen, if a narrow beam from a small circular hole is allowed to fall on a prism with an equilateral cross section, a blurred patch of colours results where the refracted light falls on the screen. By placing a narrow parallel-sided slit in front of the aperture the result is improved, and if, before the prism is placed in position, a convex lens is so adjusted as to throw a sharp image of the slit on the screen, and the prism is then placed between the lens and the screen, a much clearer spectrum is obtained. This can be still further improved by slightly rotating the prism until the position of minimum deviation is reached. The spectrum, although small, will then be pure and the spectral colours, described as red, orange, yellow, green, blue, indigo, violet, may be distinguished. Of these the red has been deflected least and the violet most.

If a second similar prism is now placed behind the first one, as shown in Fig. 70 *a*, a considerable increase both in deviation and dispersion is caused, the spectrum being more spread out. Very broad spectra for astronomical purposes are obtained by using a train of prisms. By placing between the prisms a screen of cardboard with a narrow vertical slit in it, it is possible to cut off most of the colours transmitted by the first prism so that the light falling on the second prism is of one colour only. The image on the screen is then also of one colour, and not a faint but complete range, which might be expected assuming colour to be due to a condensation caused by the prism.

Moreover if the second prism is turned so as to place its axis horizontal, and so cause a deflection at right angles to that caused by the first one, a result is obtained which once again does not bear out the old theory of colour. Instead of

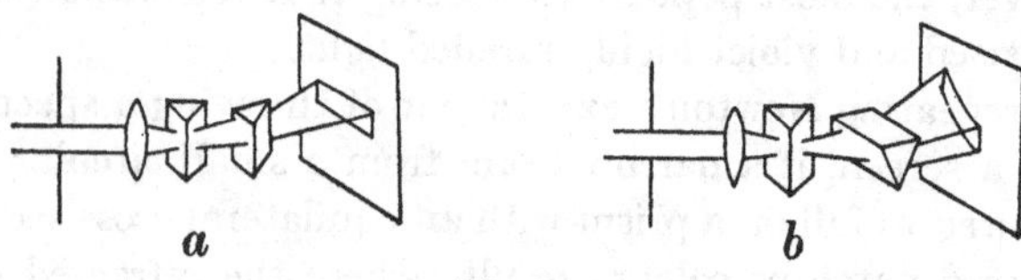

Fig. 70

the whole spectrum being shifted up, the violet end is tilted up most and the red end least, the band of colours lying obliquely (Fig. 70 *b*). This is precisely what one might expect assuming that the prism separates the colours by virtue of their different refrangibilities. Violet light is bent aside more than blue, blue more than green, and so forth, red being bent least. If we recall what it is that causes refraction, it appears as though different coloured lights travel at different speeds through glass. This actually is found to be true, red light having a greater velocity constant than other colours and violet a smaller one, for a great many refracting media, including glass.

Following up this line of reasoning it appears as though the function of the prism is merely to sort out the colours, and if so they must all have been present in the original white light. This can be tested if by some means we can recombine the colours of the spectrum.

There are two simple methods of doing this. One is merely to reverse the second prism we have used, and when this is done a white patch is again produced on the screen. An alternative method is to place the set of small mirrors shown in the diagram so that they spread across the spectrum, and each mirror reflects a different colour. By means of a slight shift in the positions of the mirrors the light reflected from each can be made to fall on the same spot on the screen, when once again white light is obtained.

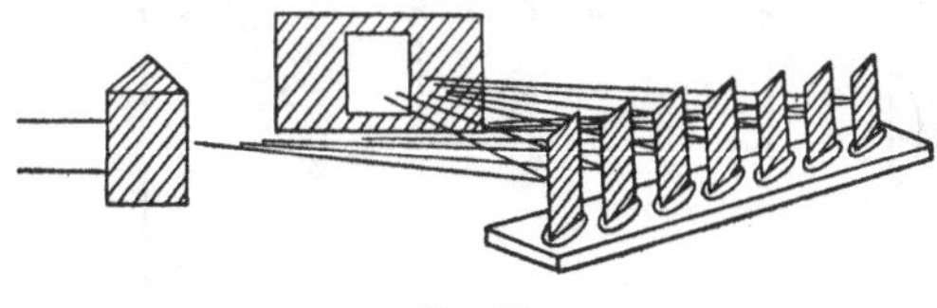

Fig. 71

For this experiment, as well as for some of those to be described later, the spectrum produced by an ordinary glass prism is much too small and crowded. A greater dispersion, or sorting out of the colours into a longer band, can be obtained by means of a hollow glass prism filled with carbon disulphide. Far superior to any ordinary prism method, however, is the means devised by Dr Hartridge, whereby a pure spectrum of any size up to 6 ft. in length can be obtained on a screen. The following description is reproduced by kind permission of the editor of the *School Science Review*:

An arc lamp, with carbons at right angles and their tips almost in line, is a convenient source of light to use, but I have found a ten amp. Triumph Focuslite gas-filled filament lamp very satisfactory, and on alternating current it is less troublesome than an arc. An arc on A.C. tends to rotate about the carbons and to get out of position.

The light is concentrated with a small condenser, *A*, on to a slit, *B*, measuring 5 mm. by 1 mm. A suitable condenser is about 5·5 cm. diameter and 6 cm. focal length. The slit can be made from razor blades, mounted in a metal frame. If an arc lamp is used, it is a slight advantage to place a six diopter cylinder lens between the condenser and the slit, to spread the light in a vertical direction over the slit. The lens, *C*, is achromatic, 3·8 cm. diameter, focal length—5·5 in. *D* is a diaphragm to stop stray light, *E* another achromatic lens, 5·4 cm. diameter, focal length—10 in., and *F* another diaphragm. *G* is a right-angled prism, the hypotenuse measuring 7·5 × 6 cm. On the hypotenuse is cemented, with Canada balsam dissolved in xylol, a diffraction grating. This

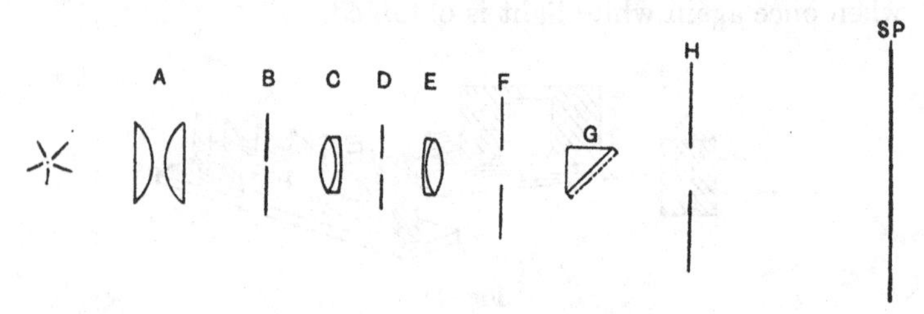

Fig. 72

is a Browning replica of a Rowland diffraction grating, 14,438 lines per inch and mounted on a rectangular block of glass. *H* is a screen with a rectangular hole in it to cut out stray light.

To adjust the apparatus, the prism with its grating is removed. The lamp is adjusted to illuminate the slit as brightly as possible, and the lenses *C* and *E* are moved up and down the optical bench to give a clear real image of the slit on the screen. The prism is then put in the position shown, and a splendid spectrum results at *SP*.*

Having obtained a spectrum with a prism and recombined the colours with a second prism, it is instructive to push a straight-edged piece of cardboard into the spectrum that exists between the two prisms. As soon as the card reaches the edge of the light, so as to become tinged with colour, the

* E. H. Duckworth in *S.S.R.* No. 20, June 1924.

recombined patch also becomes coloured, but with a colour different from that on the card. By subtracting one colour from white light we have produced another. Colour in fact is something less than white light. We produce colour by suppressing part of what was there before.

This can be demonstrated in a very striking manner by obtaining a large pure spectrum on a screen and then placing variously coloured glasses or pieces of gelatine, and also bottles containing coloured liquids, in the path of the beam of light falling on the prism. Using a piece of good ruby glass we produce a large black area covering all the violet, blue and green, and most of the yellow. Cobalt glass extinguishes the yellow completely; and a solution of permanganate of potash cuts out the whole of the midde of the spectrum and allows the ends to pass. The absorption spectra of various solutions are quite characteristic, and it is interesting to compare different liquids which, though similar in colour to the eye, prove in the spectrum to be transmitting very different mixtures of coloured lights to produce the same effect.

Just as it is possible to produce, say, orange light by quite different mixtures of colour, so it is possible to obtain the sensation of white light in several different ways. In the path of the broad spectrum, produced by the method described, a series of cards bearing slits separated by different distances can be placed, thus allowing different parts of the spectrum to pass. The transmitted light can be brought to a focus using either a large cylindrical lens, or as a substitute a circular crystallising dish full of water.

When blue and yellow light only are allowed to pass, a patch of white is produced where they combine. Those with a knowledge of colour mixing might expect green to result, but not white. A mixture of green and red light also produces white. Pairs of colours which, when mixed, produce white in this manner are said to be *complementary*, and a table of complementary colours is given below:

Crimson	**Moss green**
Scarlet	**Peacock blue**
Orange	**Turquoise blue**
Yellow	**Blue**
Primrose	**Violet**
Green-yellow	**Purple**

Just as white light can be produced by a mixture of complementary colours, so by subtracting one of these from white light we can produce the other. This can be shown very prettily, using the seven mirrors. The spectrum is first recombined to give a white reflected patch, and then different mirrors are turned aside in turn. The two patches produced exhibit complementary colouring.

When a compound of sodium is held in the bunsen flame it colours the flame a brilliant yellow. An excellent arrangement for producing this effect is shown in the diagram. A loop of thick iron wire is wound round with asbestos fibre and then dipped in water-glass. It is supported in a horizontal position over a bunsen as shown, and gives a good yellow light, while at the same time avoiding the mess that is inevitable when common salt is held in the flame. The spectrum of the sodium flame shows that it consists solely of yellow light, and so gives a single yellow line. Actually there are two lines, but they are very close together, and unless very special means are used for separating them they overlap. Such light is called *monochromatic*. Viewed in sodium light, coloured pictures have a curious appearance. All colours appear as greys or blacks with the exception of yellow, which looks very bright. The livid appearance of the human countenance is ghastly.

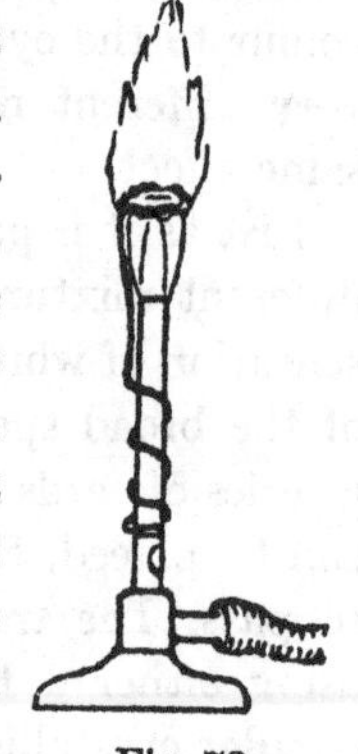

Fig. 73

We may extend this experiment and pass coloured pictures through all the colours of the spectrum in turn. The apparent changes are very instructive, and an interesting application of this effect was made recently in stage lighting, when the scenes were changed merely by changing the colour of the illumination.

Now the light from sodium is a pure yellow. It is possible to produce a yellow light by taking its complementary colour, blue, out of ordinary sunlight. This yellow, however, contains all the spectral colours with the exception of blue. Hence we must distinguish between the actual colour present and the sensation produced whereby we recognise it. So far as can be judged there are three different colour sensations associated with the eye, these being red, yellowish green and blue-violet. Each of these colours excites one sensation and is known as a *primary colour*. Any other colour excites more than one. A pure yellow excites both the red and the green sensations, and a blue the green and violet. Thus yellow light together with blue light excites all three sensations and so produces the effect of whiteness. These primary colour sensations must not be confused with the so-called "primary colours"—red, yellow and blue—of a paint-box from which other coloured pigments may be obtained by mixing. Blue paint mixed with yellow paint gives green paint. On the other hand, blue light mixed with yellow light gives white light. This sounds like a flat contradiction. The explanation depends on the essential difference between a coloured light and a pigment that is similarly coloured.

Red paint, for instance, is red because it absorbs all the other colours of white light and reflects back red only. If supplied with light from which the red has been removed, red paint does not reflect any light at all and consequently appears black.

Blue paint absorbs yellow and red light but reflects blue

and green. Yellow paint absorbs blue and violet light and gives back yellow and green. Thus a mixture of the two suppresses both blue and yellow, but the green common to both is reflected.

The pigments used in water-colour painting are semi-transparent. White light is reflected by the paper at the back, and from this reflected light the pigments absorb the appropriate colours. Hence brilliance is obtained by thin washes of colour, and in this respect there is a marked contrast to oil painting where the "colours" have body and are themselves the reflectors.

Actually very few pigments are pure in the sense of reflecting monochromatic light. Thus by viewing a coloured object in different lights it is possible to get a series of different combinations of colour reflected, according to the method of illumination. The red of flowers is enhanced by artificial light in which there is a preponderance of red: and it is a frequent embarrassment that materials which "match" perfectly by artificial light show distinct hues in the more searching and complex light of day.

There are two other methods of mixing colours. One consists of painting sectors of cardboard disks with different pigments, and then revolving the disks at high speed. The eye is unable to follow the movement, and the resulting sensation is a blend of those produced by the separate colours. As in this case the light is reflected from pigments, there is a tendency for a greyish hue to be present.

A much more satisfactory but very complex instrument is the colour-box designed and made by Clerk Maxwell. It consists of a train of prisms (see Fig. 74) so arranged that light entering at A is split up into a long spectrum, the position of which is indicated by the letters V and R. If now the place occupied by the spectrum is covered by a series of shutters, and then white light is admitted at, say, R, only red light will reach *A*, the other colours being refracted too much. In a similar way if *V* is illuminated with white light, only violet light will reach *A*, since the other colours are not refracted enough. Thus, by admitting white light through appropriate slits, the effect of all possible combinations of colours can be seen by an observer looking through the telescope. The instrument called the spectro-heliograph, by means of which the sun is photographed daily in any selected monochromatic light, is very similar in construction.

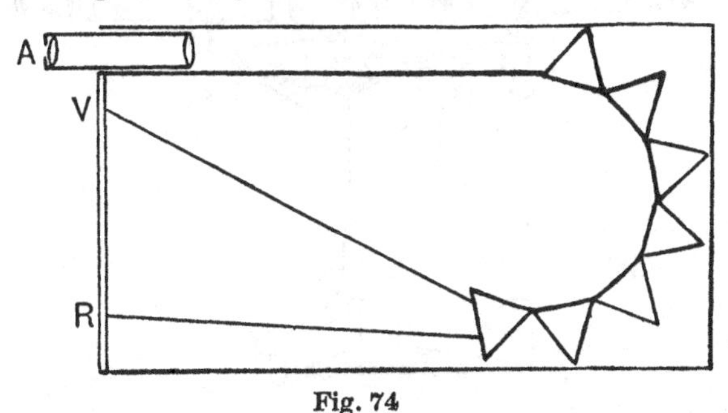

Fig. 74

When it is desired to examine the spectrum in greater detail an instrument called the spectroscope is employed (see Fig. 75). There are three parts, the collimator, table and telescope. The collimator consists of a tube at one end of which is an adjustable slit, *H*, placed at the principal focus of a lens, *G*, so that light from the slit leaves the lens in a parallel beam. It then falls on the prism, *K*, placed on a table which can be rotated until the position of minimum deviation is obtained. The spectrum so formed is viewed through the telescope, *DE*, which is previously adjusted to pick up parallel light.

The light from a sodium flame gives a single yellow line, that may appear double with a very good instrument. From a carbon arc lamp a continuous spectrum is seen. When electric discharges are sent through rarefied gases, the light produced gives a discontinuous spectrum consisting of a series of bright bands, the number and position of which depend upon the gas used.

For obtaining a solar spectrum it is convenient to employ a device similar to that used at observatories. A convex mirror is placed in the sun's rays, and the small virtual image in the mirror is used as the source for illuminating the slit.

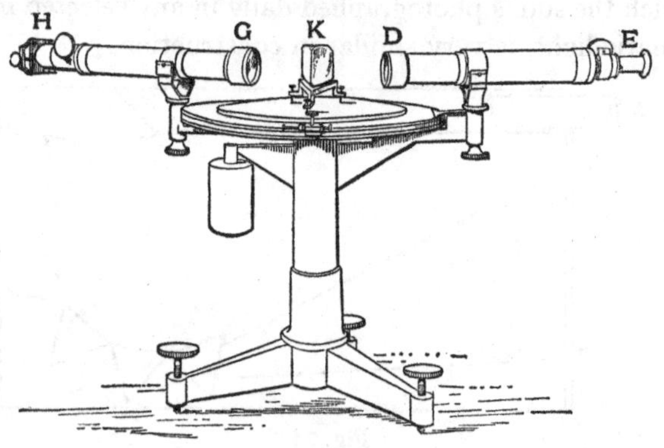

Fig. 75

The spectrum so produced is seen to be crossed by a number of black lines, named the *Fraunhofer lines*, after the man who first pointed them out. When a prism is used several dozen lines are visible, but the much more widely dispersed spectrum that can be obtained by means of a diffraction grating shows that there are many thousands.

The curious thing is that these lines correspond exactly in position with the bright lines produced when electric discharges are sent through various gases. A simple experiment will help us to understand why they are dark and not bright.

The spectroscope is set up and the yellow line of sodium is found, using a burner fitted up in the ordinary way. An arc lamp is then lit exactly behind the burner and in a line with the slit. Light from the arc must pass through the yellow flame to reach the slit, and with a few adjustments a spectrum is obtained with a dark line in the yellow corresponding to the position of the yellow line of sodium.

The accepted explanation is as follows. In order to produce yellow light the atoms of sodium in the flame are vibrating with a certain frequency. When the much more intense waves from the very much hotter arc pass through the flame, those having the same frequency as the sodium atoms are very largely absorbed by them, and the light passing through is deficient in these particular waves as a result.

The Fraunhofer lines are produced in the same way. The sun must consist of an intensely hot inner portion, the *Photosphere*, surrounded by a relatively cool gaseous envelope, the *Chromosphere*, in which the different elements are present as vapours. These rob the original light of their characteristic vibrations and so produce the black lines.

The light from the fixed stars, which are also suns, exhibit similar lines, and from a study of these lines it is possible to tell which elements are present in the particular star. The comparatively new science of astrophysics is largely concerned with mapping and interpreting the detailed spectra of the stars.

Huyghens' construction, as modified by Ampère, helps us to follow how the sorting process due to a prism takes place, and gives us the direction of the different colours produced by refraction. The construction is similar to that employed on p. 60 for refraction at a plane surface, but two circles are drawn around A corresponding to the different values of h for red and violet light (see Fig. 76). For crown glass h using red light is 0·664 and with violet light it is 0·654. AB is the advancing wave and XY the surface. B has still to travel BD in air. With A as centre and radii 0·664 BD and 0·654 BD circles are drawn, and the tangents DR, DV constructed.

Then *RD* is the red wave-front travelling along *AR*, and *VD* is the violet travelling along *AV*.

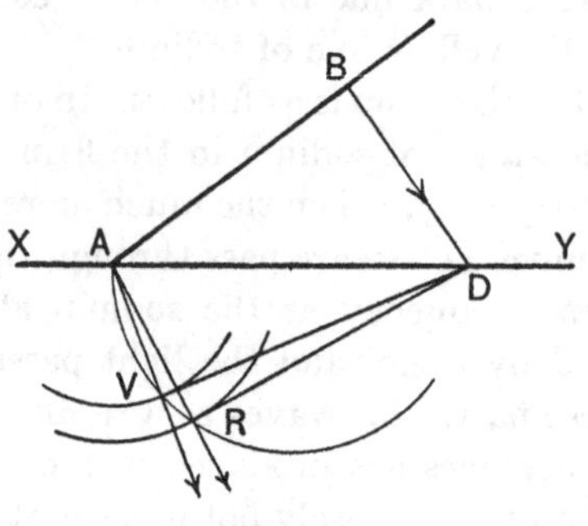

Fig. 76

Owing to the two values for h it is obvious that the principal focus of a lens cannot be at the same point for red and blue light. Fig. 77 indicates that when light from a distance is brought to a focus for red light it will be surrounded by a violet ring, and vice versa. This defect is called *chromatic aberration* and for a long time it was supposed to be unavoidable. A discovery by Dollond showed that it is possible to obtain refraction without dispersion by means of two kinds of glass.

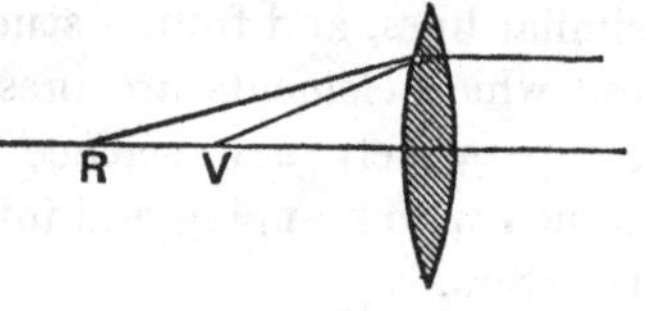

Fig. 77

The dispersive power of a medium, *i.e.* the degree of separation of the colours in a spectrum obtained using a prism of the medium, is a measure of the *difference* between its velocity constants for the different colours. Flint glass has a much greater dispersive power than has crown glass,* and hence to obtain a spectrum equal in width to that from a

* For flint glass h is 0·61 using red light and 0·597 using violet.

given crown glass prism, a prism of flint glass with a considerably smaller angle can be employed.

Although by this means an equal *separation* of colours is obtained, the average deviation due to the thinner prism will be less. Hence, if a spectrum is produced by means of the crown glass prism and recombined by the flint glass prism

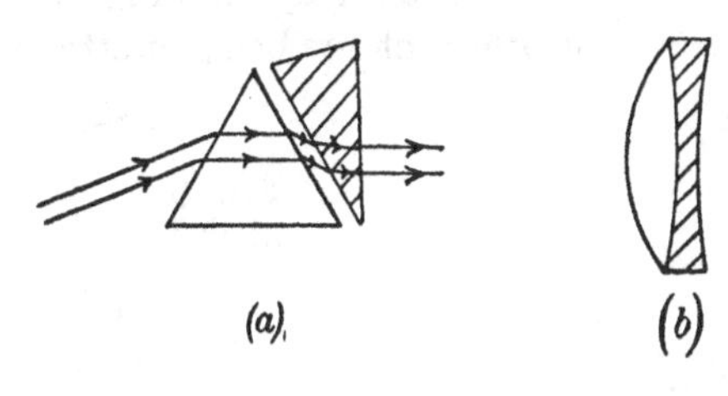

Fig. 78

of equal dispersive power, white light results but it is deviated from its original course (see Fig. 78). In a similar way, a compound lens such as that shown in the figure brings all the colours to a focus at the same point. The name of such an arrangement is an achromatic combination.

The rainbow which appears when sunlight shines on falling rain affords the best known example of colours formed by refraction, and to Marco Antonio de Dominis (1566–1624) is due the credit of first describing how the bow is caused. The path of light in a sphere for a certain direction of incident light is shown in Fig. 79 (*a*). It is refracted on entering, reflected at the back and refracted again on leaving. Dispersion of colours will occur at each refraction and the violet, being refracted most, will be at the top of the emergent bundle of colours.

Now this same state of affairs will occur in every drop of rain, the violet ray, for instance, making the same angle with the incident ray in each drop. Fig. 79 (*b*) illustrates the direction of the violet, yellow and red rays from three spheres, the dispersion being greatly exaggerated. Owing to the fact that the incident light from the sun is parallel, all the emergent violet rays V_1, V_2, V_3 will be parallel, and so will those of other colours. As a result, an observer will see red light coming from above, yellow from underneath that and violet from lower still, as is indicated by showing R_1, Y_2 and V_3 reaching the eye, the other colours being omitted for the sake

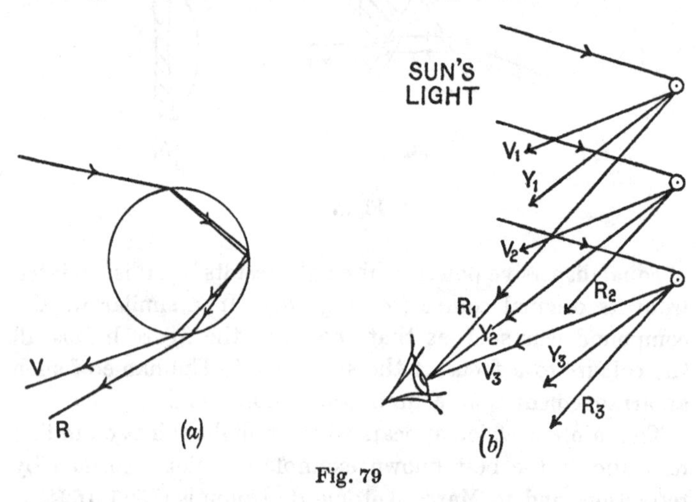

Fig. 79

of simplicity. Moreover refraction does not take place merely in one plane in the drop of water but in three dimensions, and the light coming away from the sphere is in the form of part of the surface of a cone with the drop at its apex. Thus all the drops from which a particular colour can reach the eye lie on a circle, and so the well-known arch, with the violet on the inside and the red outside, appears against the falling shower.

The formation of colours inside a spherical drop can be shown by inverting a small spherical bulb full of water, and directing into it a narrow beam of light from the lantern. If a screen is placed round the nozzle of the lantern as shown in Fig. 80 a coloured arc of a circle appears on the screen.

Exploration of the regions beyond either end of the visible spectrum has revealed that the visible part is only a small fraction of the total spectrum. If a thermometer bulb, coated with dull black, is placed just beyond the visible red, a rise in temperature is at once recorded showing the presence of the heat waves of the "infra-red". A photographic plate

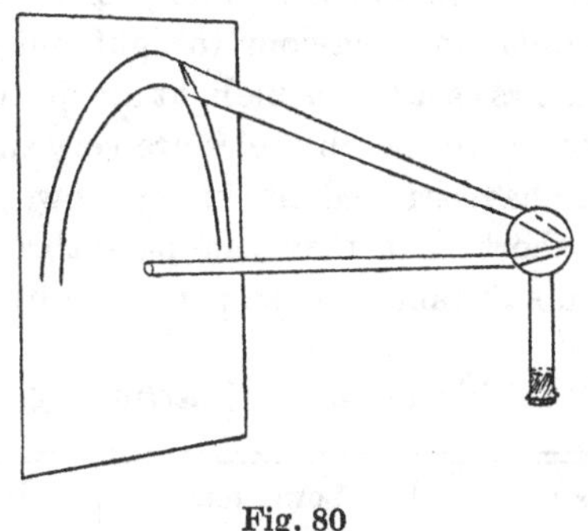

Fig. 80

reveals that beyond the violet there is a region of invisible but chemically active "ultra-violet" waves, which can be rendered visible using a fluorescing screen such as X-ray workers employ. When glass prisms and lenses are used, much of the infra-red and ultra-violet parts are absorbed, but when quartz lenses and prisms are employed the vast extent of the ultra-violet can be gauged. For infra-red waves it is necessary to use apparatus of rock-salt.

Modern research has revealed that visible and invisible light, so-called Radiant Heat and the Hertzian waves employed in wireless telegraphy and telephony, are all essentially similar, all being electro-magnetic in nature and differing only in the length of the waves. Moreover the X-ray is also a form of light whose wave-length is exceedingly minute, being only from 1/40 to 1/4000 of the wave-length of yellow light. The table showing the ranges of wave-lengths reveals how comparatively insignificant is the short "wave-band" that can affect our eyes. The peculiarities of light are all attributable to its very short wave-lengths. The long "wireless" waves do not form shadows and cannot even be persuaded to travel in parallel beams. Quite recently much work has been done with short waves and so "beam" wireless has been evolved with much economy of energy. Radiant heat is very similar in behaviour to light but spreads more easily. The shadows cast by visible light are quite sharp except at the edges or where the obstacles are very small, when the light is found to have strayed into the shadow. Finally X-ray waves are so short that they can pass through the very atoms of a solid substance and so penetrate it.

TABLE TWO. *Wave-lengths of electro-magnetic waves*

Type of wave	Lower limit	Upper limit
Gamma rays	1×10^{-10} cm.	$1{\cdot}4 \times 10^{-8}$ cm.
X-rays	6×10^{-10} "	1×10^{-6} "
Ultra-violet waves	$1{\cdot}4 \times 10^{-6}$ "	4×10^{-5} "
Visible light	4×10^{-5} "	8×10^{-5} "
Infra-red radiations	8×10^{-5} "	4×10^{-2} "
Short Hertzian waves	1×10^{-2} "	1×10^{3} "
"Wireless" waves	5×10^{2} "	3×10^{6} "
Longer waves	3×10^{6} "	—

The method of rendering beams of light visible by smoke clouds has been mentioned. When light is scattered in this way, the short blue rays are scattered far more than the longer red ones. For this reason cigarette smoke and water into which a little milk has been stirred both have a distinct bluish colour. Moreover it is to this scattering that we owe the glorious colour of the sky, and of the setting sun.

Consider what is happening when the sun is low on the horizon. As the originally white light passes through the atmosphere the blue waves are scattered, and when a considerable thickness of air has to be traversed the scattering is fairly complete. Hence the red light reaches us direct from the sun, while from all other directions we see the glorious blues and greens that have been scattered. Were there no atmosphere there would be no blue sky.

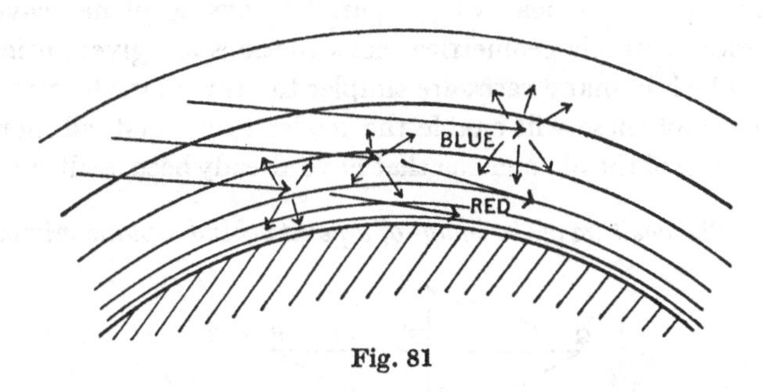

Fig. 81

The red appearance of the sun on a foggy day is due to a similar cause. This property of the longer waves has been utilised in the design of fog-piercing motor headlights, which have red filters to absorb the blue waves that cause the "glare" in foggy weather. A still more surprising development is promised in the near future with the perfection of fog-piercing rays of long infra-red light, which can be rendered visible by special means.

Chapter Eleven

GEOMETRICAL OPTICS

FOR purposes of revision, the concept of a "ray" of light is useful. A ray is merely a line showing the direction of movement of a wave-front, and under no circumstances is it possible to obtain a geometrical line of light. Converging rays represent a hollow spherical advancing wave; diverging rays a diverging spherical wave; parallel rays a plane wave. A series of simple geometrical constructions are given, using rays which in many cases are simpler to draw than the wave-fronts, and these will enable the reader to understand more fully some of the phenomena that have already been dealt with.

1. *The position of the image of a point behind a plane mirror.*

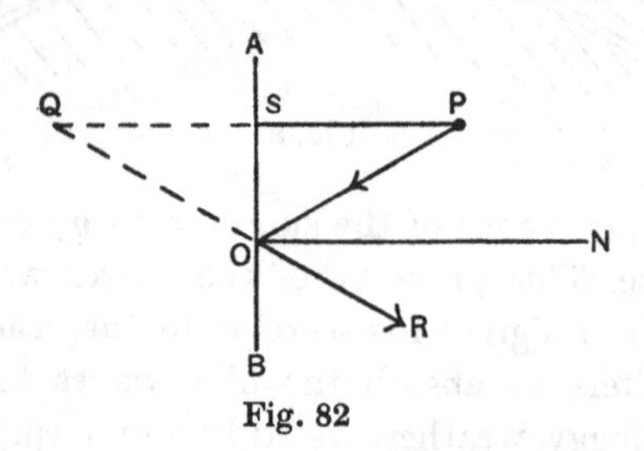

Fig. 82

AB is the mirror surface, *P* the luminous point. Draw a line through *P* perpendicular to *AB*. This represents light falling normally on the mirror, and under these circumstances it is reflected back along its own path.

Draw any other line *PO*. Construct *ON* perpendicular to the mirror and draw the line *OR* so that $\angle RON = \angle PON$. Produce *RO* until it meets *PS* produced in *Q*.

Q is the image of *P*.

Proof. From the construction the triangles *PSO*, *QSO* are congruent. (Right angle, one other angle, common side.)

Therefore $QS = PS$.

This fulfils the necessary conditions:

(*a*) The angle of incidence = angle of reflection.

(*b*) The image is on the same normal as the object and is as far behind the mirror as the object is in front.

2. *To trace the rays by which a point image is seen.*

This is the simplest case of a number of problems, and the method of solution is always the same.

Draw in the object, mirror, image and eye. From either side of the pupil of the eye draw straight lines to the image. Part of these lines will lie behind the mirror and this part should be dotted. These lines are the boundaries of the rays that enter the eye, and must have originated at the object. Join the points where they touch the mirror surface to the object. Mark arrows on the lines to show the direction in which the light is travelling. The dotted portions can have no real existence, and these are termed *imaginary rays.*

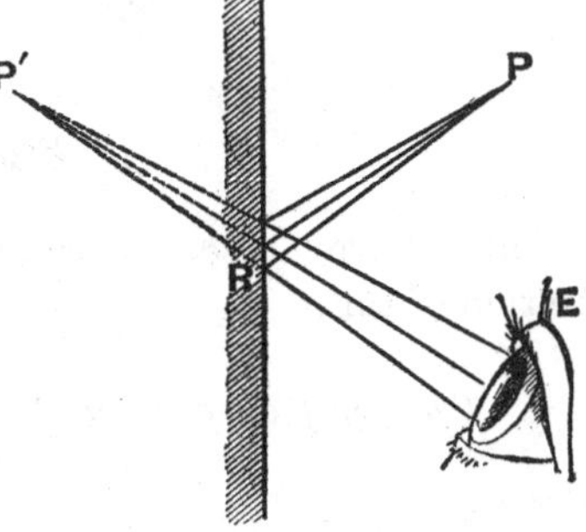

Fig. 83

3. *To draw the rays whereby a large image is seen.*

The method is precisely similar to that already described, a point at each end of the object being treated separately.

It will be noted in the drawing that the end of the image nearer to the observer corresponds to the end of the object that is farther away. This turning-round effect is called *lateral inversion*. The well-known trick of writing on paper, blotting the ink while wet and then holding up the blotting-paper

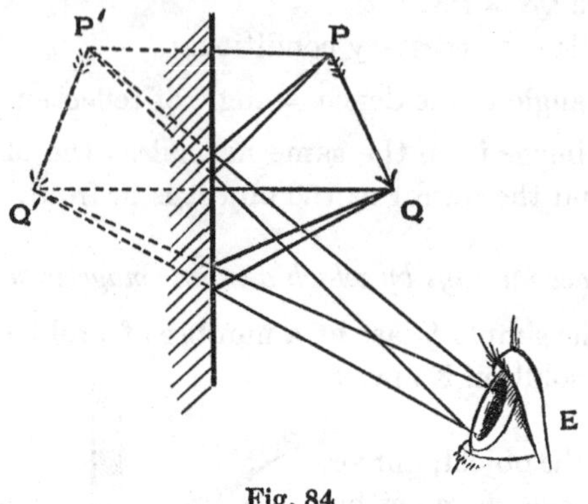

Fig. 84

before a mirror so as to read what was written, is an illustration of this.

4. *The rays from a point placed between two parallel mirrors.*

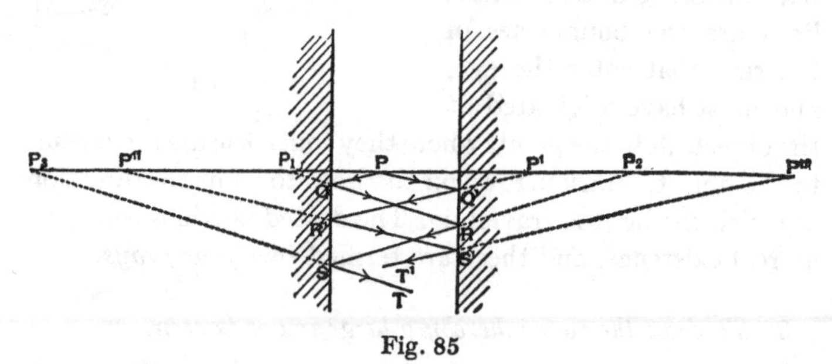

Fig. 85

The normal is drawn through the point and any two rays PQ, PQ' are drawn to either mirror. The reflected rays QR and $Q'R'$ are drawn and produced back giving P_1 and P', the images of the first order. P_2 and P'' are second order images formed by two reflections, P_3 and P''' are third order images and so forth. If the mirrors are exactly parallel an infinite number of images can be formed. Large mirrors placed on opposite walls of a room give an effect of an endless corridor stretching away on either hand until it vanishes in the distance.

5. *The position of the images of an object in two mirrors inclined at* 60°.

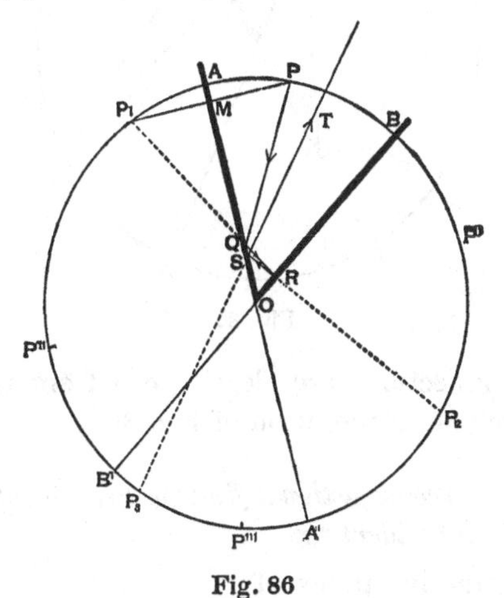

Fig. 86

Five images are formed, their positions being determined by constructing the normals. There are two first order images, two second order and one third order image. This lends itself to various optical illusions. For instance if a triangular slice of cake is placed between the two mirrors, a whole cake appears, due to the continuity of the images. The kaleidoscope and other toys of a similar nature depend for their working on the symmetry of the patterns formed by inclined mirrors.

6. *The rays whereby the third order image is seen between the two mirrors at 60°.*

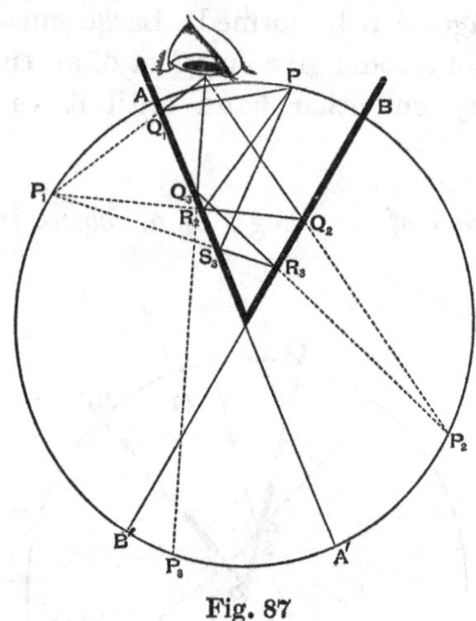

Fig. 87

The three reflections are clearly seen from the diagram, which is merely an elaboration of Fig. 86.

7. *Geometrical construction to find the direction of the refracted ray from a given incident ray.*

Let the refractive index of the medium $\mu = x/y$. *BA* is the refracting surface and *PR* any incident ray. With centre *R* draw a circle with any radius cutting *PR* in *P*. Drop a perpendicular from *P* to *BA*. Draw also the normal through *R*. Divide the distance *KR* into x parts.

Mark off $LR = y$ parts.

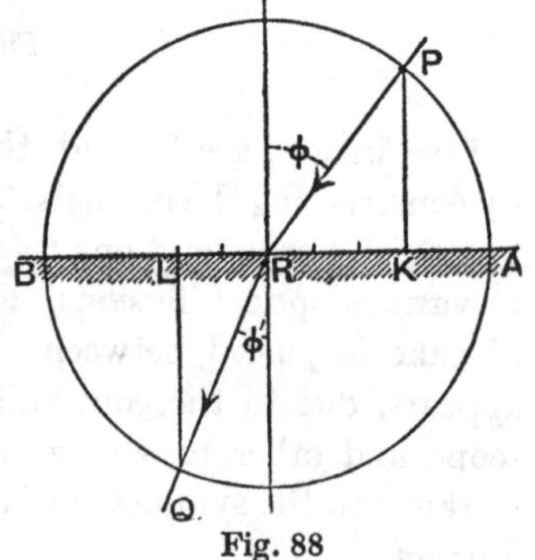

Fig. 88

Draw through L a perpendicular cutting the circle below BA in Q.

Join RQ. Then RQ is the refracted ray.

Proof. $\left.\begin{aligned} \angle RPK &= \text{angle of incidence } \phi \\ \angle RQL &= \text{angle of refraction } \phi' \end{aligned}\right\}$ see diagram.

Then $$\frac{\sin\phi}{\sin\phi'} = \frac{KR/\text{radius}}{LR/\text{radius}} = \frac{KR}{LR} = \frac{x}{y} = \mu.$$

8. *The angle of minimum deviation of a prism.*

By experiment it is found that there is only one position of the prism in which the deviation is a minimum.

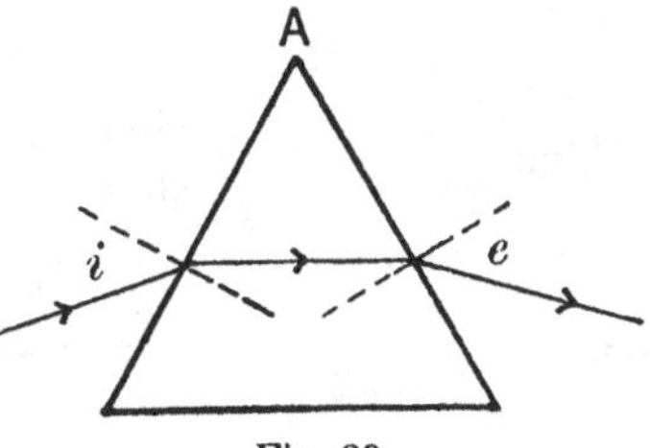

Fig. 89

Let $PQRS$ be the path of light for this position. Suppose the path reversed. The deviation is still a minimum and so the light must follow along a course precisely similar to its former one. This is only possible if the light passes through the prism symmetrically with respect to the apex A.

Thus it is seen that for minimum deviation the angle of incidence i must be equal to the angle of emergence e.

9. *The connection between the angle of minimum deviation and the angle of the prism.*

$PQRS$ is the ray that suffers minimum deviation (see Fig. 90).

Call $\angle SVW$ the angle of deviation $= D$.

From the figure $\angle i = \angle r + \angle\beta$.

And since $\angle QAR + \angle QOR = 2 \text{ rt } \angle\text{s}$

and $\angle QOT + \angle QOR = 2 \text{ rt } \angle\text{s},$

therefore $\angle QAR = \angle QOT = 2\angle r,$

i.e. $\angle r = \angle A/2.$

Further $\angle D = 2\,\angle\beta$,

therefore $\angle\beta = \angle D/2$.

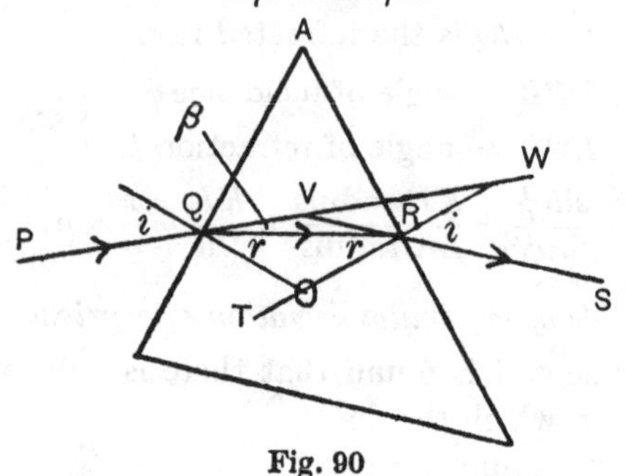

Fig. 90

Thus $$\frac{\sin i}{\sin r} = \frac{\sin \beta + r}{\sin r},$$

i.e. $$u = \frac{\sin\dfrac{A+D}{2}}{\sin\dfrac{A}{2}}.$$

10. *Refraction of light at a plane surface giving a caustic curve.*

When a number of rays are drawn from a point P below the surface of a refracting medium, and the refracted rays aA, bB, cC, etc., are drawn, these are found not to diverge from a point but to coincide along a curve as shown in the figure. The curve is called a caustic.

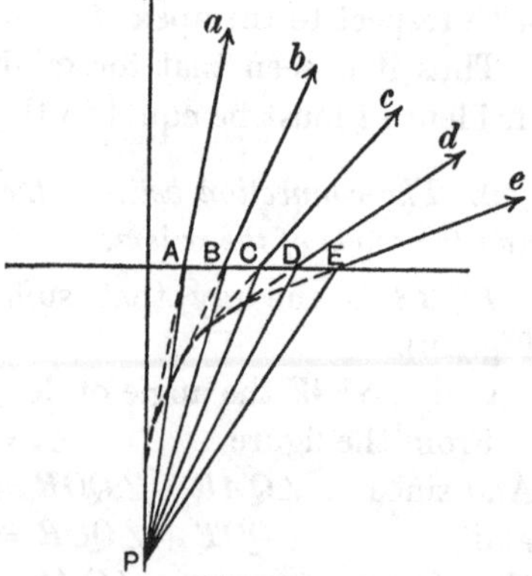

Fig. 91

The relation

$$\frac{\text{Real depth}}{\text{Apparent depth}} = \mu$$

does not hold, therefore, except for a very narrow pencil of rays emerging almost normally. This corresponds to the very small fraction of the spherical surface mentioned in establishing the relationship (see p. 62).

GRAPHICAL METHODS FOR MIRRORS AND LENSES

In dealing with mirrors and lenses we make the following assumptions:

(*a*) All rays originally parallel to the axis of the lens or mirror are altered in direction so as either to pass through the focus or to appear to diverge from an imaginary focus.

This is true only when the aperture of the lens or mirror is small, *i.e.* when its surface constitutes only a very small fraction of the surface of a sphere.

(*b*) In the case of a lens, a ray passing through the centre of the lens is unchanged in direction.

This is true when the thickness of the lens is so small that it can be neglected.

(*c*) Rays falling normally on a mirror, *i.e.* rays through the centre of curvature, are reflected back along their own path: and rays falling anywhere on the mirror are reflected back so as to make an equal angle on the other side of the radius to the point of incidence.

All these assumptions can be tested by experiment and found to hold good.

11. *Images formed by a concave mirror.*

BAC is the mirror, *O* the centre of curvature. It has been shown already (p. 63) that the principal focus *F* is midway between *O* and *A*.

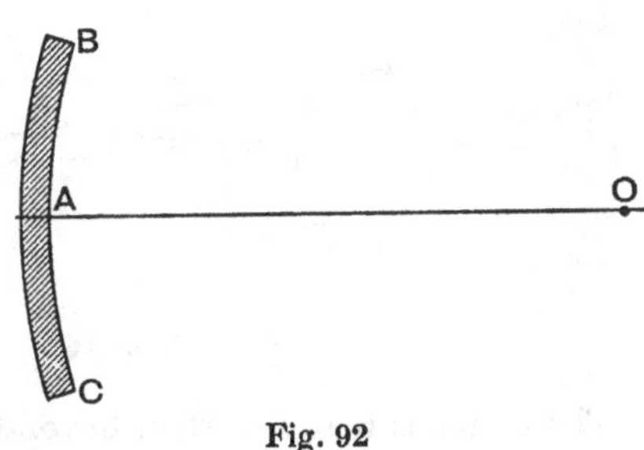

Fig. 92

Case (*a*). Object beyond *O*.

The object is shown on one side only of the axis for the sake of simplicity.

The ray *PR* parallel with the axis is reflected back through *F*.

The ray *PT* through the centre of curvature is reflected back along its own path.

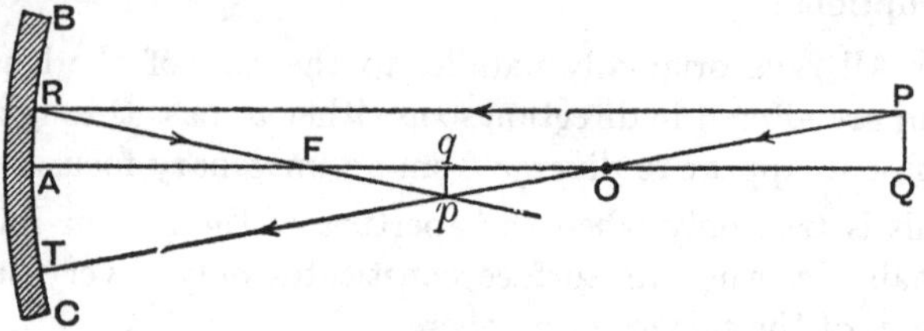

Fig. 93

These coincide at *p*, which is the image of *P*, while *q*, the image of *Q*, is at a corresponding place on the axis.

Thus the image is real, inverted, between *O* and the focus and smaller than the object.

Case (*b*). Object at *O*.

In this case by the construction the image is real, inverted, at *O* also and the same size as the object.

Case (*c*). Object between *O* and *F*.

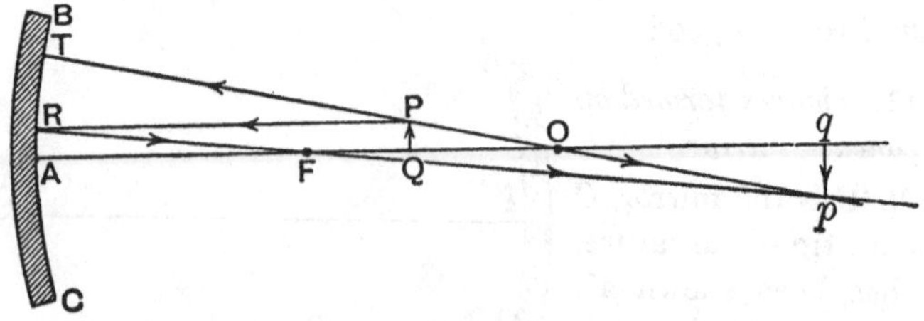

Fig. 94

The image is real, inverted, beyond *O* and magnified.

Case (*d*). When the object is at the focus the reflected light is parallel. Hence there is no image, or the image may be described as at infinity. With the object at infinity of course there is a point image at the focus.

Case (*e*). Object between *F* and the mirror.

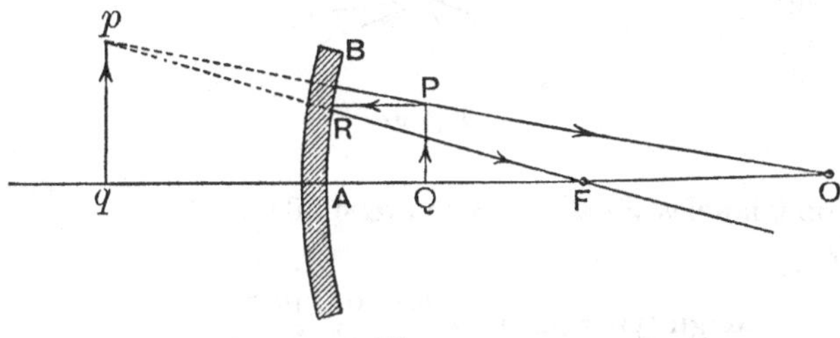

Fig. 95

This time the reflected rays are divergent. They appear to come from a point behind the mirror and so a virtual, erect, magnified image is produced behind the mirror. This case illustrates the use of concave mirrors as shaving-glasses.

12. *Images formed by a convex mirror.*

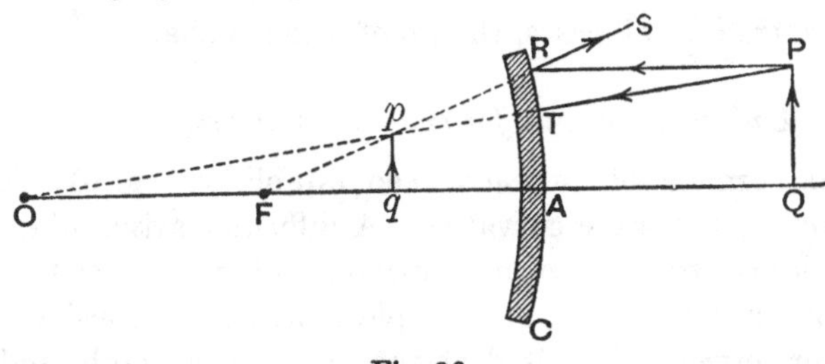

Fig. 96

The method used is the same in essentials. As however the image is always virtual, behind the mirror erect, and smaller than the object, only one position is drawn.

13. *Magnifying power of a spherical mirror.*

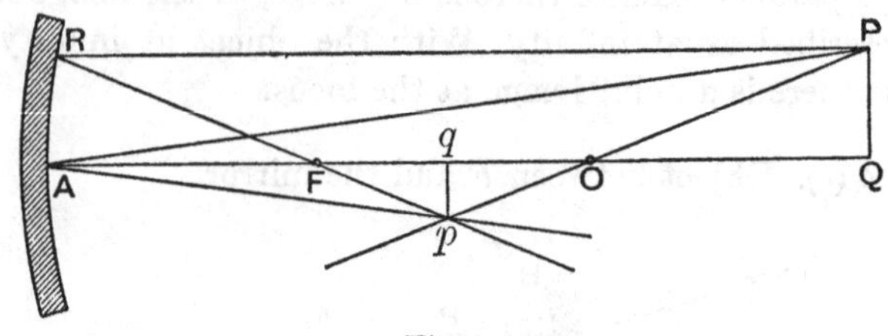

Fig. 97

The triangles PAQ and pAq are similar.

Hence

$$\text{Magnifying power} = \frac{\text{size of image}}{\text{size of object}}$$

$$= \frac{pq}{PQ}$$

$$= \frac{qA}{QA} \text{ or } \frac{v}{u}.$$

Thus $$\text{Magnification} = \frac{\text{distance of image}}{\text{distance of object}}.$$

This is true for all cases, the proof being similar.

14. *Universal formula for spherical mirrors.*

This formula has already been established (p. 64), using the concept of wave curvature. A difficulty arises when we consider the problem, using rays, concerning the signs to allot to the various terms. The simplest plan is to consider the lens or mirror as situated at the origin of a graph, and to place the object always on the right. Then distances measured to the right of the lens or mirror are positive, and those to the left negative. Unfortunately the signs obtained in this manner are the exact opposite of those corresponding to the convention of calling a convex lens positive and a concave lens negative (see p. 65). Since it affords a simple method of solving an otherwise difficult point, this plan will be adopted for the following proofs.

Consider first the case shown in Fig. 97.

By similar triangles

$$\frac{PQ}{pq}=\frac{QA}{qA}$$

and

$$\frac{RA}{pq}=\frac{AF}{qF}.$$

If the aperture is small (a necessary condition in all such cases), RA is equal to PQ.

Hence

$$\frac{QA}{qA}=\frac{AF}{qF}.$$

Substituting values of u, v and f,

$$\frac{u}{v}=\frac{f}{v-f},$$

whence

$$uv-uf=vf.$$

Dividing through by uvf we get

$$\frac{1}{f}-\frac{1}{v}=\frac{1}{u}$$

and

$$\frac{1}{v}+\frac{1}{u}=\frac{1}{f}.$$

The student should establish the formula for a convex mirror by similar means, care being taken to allot the correct signs to u, v and f.

15. *Caustic formed by reflection at a concave mirror of large aperture.*

When the aperture is large, rays originally parallel to the axis do not all converge on the principal focus. If a careful drawing is made, and reflected rays are drawn making equal angles with radii to those made by the incident rays, it is seen that the reflected rays coincide along a caustic curve similar to that shown in Fig. 98. The coincidence gives the

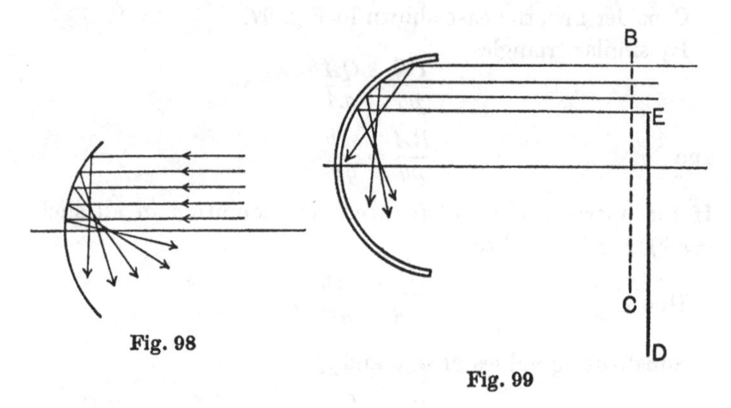

Fig. 98

Fig. 99

appearance of a curved line of light, and may be seen on the surface of a cup nearly full of milk standing in the bright light from a window. A very beautiful method of showing the formation of a caustic is described by Mr F. A. Meier in the *School Science Review*, No. 29:

Take a piece of cylindrical mirror strip. Put a *vertical* half watt filament about 20 inches from the mirror and 4 inches above the table, and a comb at *BC*. Slowly move a card *ED* across the comb so as to cut out the light. The gradual building up of a caustic is beautifully shown. A darkish room is essential (see Fig. 99).

Note. A 4 volt Midget Ediswan bulb which has a fine straight special filament, or an ordinary 60 watt horseshoe filament bulb, used sideways, gives good results.

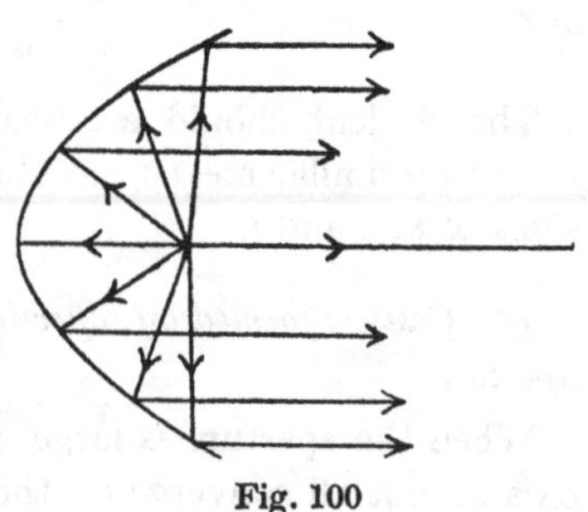

Fig. 100

In consequence of this defect of spherical aberration, whenever a mirror of large aperture is required it is not spherical but paraboloid. Such mirrors are used in reflecting telescopes (see p. 129) and as reflectors for motor headlights and for searchlights, when it is desired to throw back the light from a point source as a parallel beam (Fig. 100).

16. *Position of image formed by a convex lens.*

In the diagram *F*, *F′* represent the principal foci, *PQ* the object and *pq* the image. 2*F* is a point twice as far from the lens as is the focus.

Case (*a*). Object beyond 2*F*.

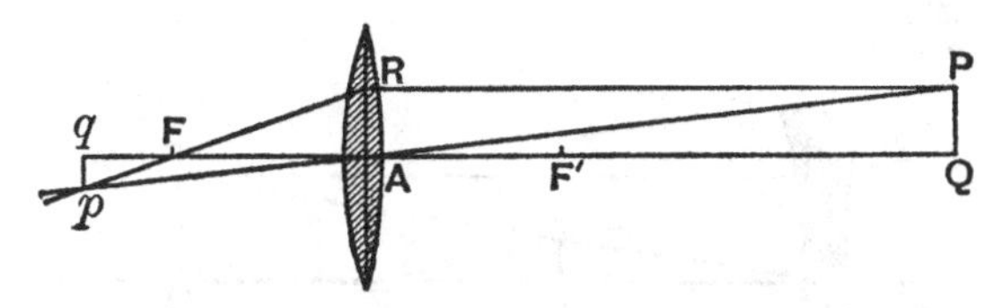

Fig. 101

The rays used are *PR*, parallel to the axis, which after refraction passes through the focus, and *PA* to the centre of the lens, which does not change in direction. In this case the image is real, inverted, between *F* and 2*F* and smaller than the object. All the other cases can be found in the same way and the student should actually draw all the possible cases.

Case (*b*). Object at 2*F*. Image also at 2*F*: real, inverted, same size as object.

Case (*c*). Object at *F*. Image at infinity.

Case (*d*). Object between *F* and lens. Image virtual, erect, magnified and the same side of the lens as the object.

17. *Image formed by a concave lens.*

The method of construction is similar to the foregoing.

In all possible cases the image is virtual, erect, smaller than the object and in front of the lens (see Fig. 102 *b*).

All cases should be drawn.

18. *Magnifying power of a lens.*

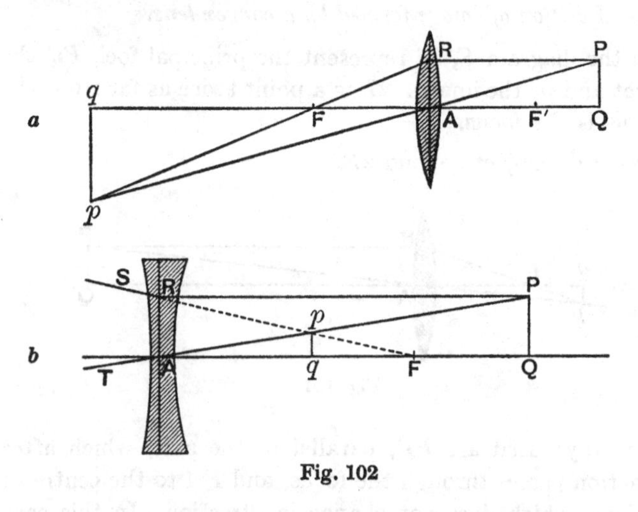

Fig. 102

From Figs. 102 *a* and 102 *b* it can be seen that, using the properties of similar triangles,

$$\text{Magnification} = \frac{v}{u} \text{ or } \frac{\text{distance of image}}{\text{distance of object}}.$$

19. *Universal formula for a lens.*

The establishment of this formula, using wave-curvatures, is given on p. 64. In Fig. 102 *a* it can be seen that, by similar triangles,

$$\frac{PQ}{pq} = \frac{QA}{qA} = \frac{AF}{qF}.$$

Substituting u, v and f, with appropriate signs,

$$\frac{u}{-v} = \frac{-f}{-v+f},$$

whence $$-uv + uf = vf.$$

Therefore $$-\frac{1}{f} + \frac{1}{v} = \frac{1}{u}$$

and $$\frac{1}{v} - \frac{1}{u} = \frac{1}{f}.$$

The proof for the concave lens is similar, but the distances u, v and f are all positive.

20. *Formula connecting focal length, radii of curvature and refractive index.*

The corresponding proof, using waves, is given on p. 72.

(*a*) Refraction at one spherical surface.

In order to simplify the question of signs the surface taken is concave. Q is a point from which light is emitted, q is its image formed by refraction and O the centre of curvature of the surface.

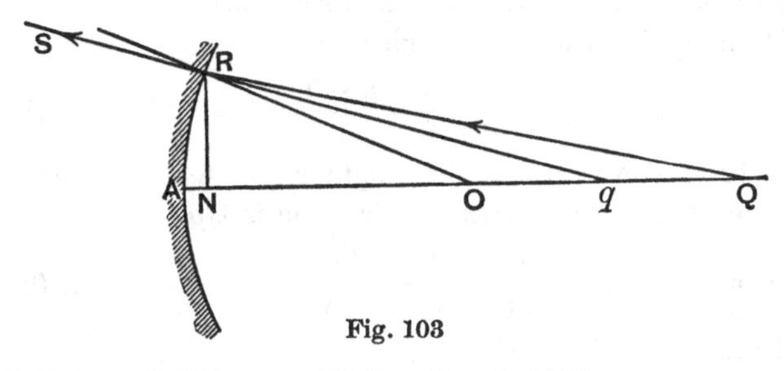

Fig. 103

Call the $\angle RON = \alpha$, $\angle RqO = \beta$ and $\angle RQq = \gamma$.

Now $\angle QRO$ is the angle of incidence, i,

and $\angle qRO$ = the angle of refraction, r.

But $\angle QRO = (\alpha - \gamma) = i$

and $\angle qRO = (\alpha - \beta) = r$.

By Snell's Law $\sin i = \mu \sin r,$

i.e. $\sin(\alpha - \gamma) = \mu \sin(\alpha - \beta).$

Provided the aperture RA is small the values of the sines are the same, to a first approximation, as the values of the angles in circular measure.

Thus $\alpha - \gamma = \mu(\alpha - \beta),$

i.e. $$\frac{AR}{r} - \frac{AR}{u} = \mu\left\{\frac{AR}{r} - \frac{AR}{v}\right\}.$$

Substituting with u, v and r we get

$$\frac{1}{r} - \frac{1}{u} = \frac{\mu}{r} - \frac{\mu}{v}$$

and $$\frac{\mu}{v} - \frac{1}{u} = \frac{\mu - 1}{r}.$$

(*b*) For a complete lens.

Call the distance of the object u and of the image formed by refraction at the first surface v'. This image serves as object for the second refraction, giving rise to the final image at a distance v. Call the two radii r and s.

At the first surface (air to glass)

$$\frac{\mu}{v'} - \frac{1}{u} = \frac{\mu - 1}{r} \qquad \text{......(1).}$$

At the second surface, since light is passing from glass to air, the value of the index of refraction is $1/\mu$.

Then $$\frac{1/\mu}{v} - \frac{1}{v'} = \frac{1/\mu - 1}{s} \qquad \text{......(2).}$$

Multiply equation (2) by μ,

$$\frac{1}{v} - \frac{\mu}{v'} = \frac{1 - \mu}{s} \qquad \text{......(3).}$$

Add equations (1) and (3),

$$\frac{1}{v} - \frac{1}{u} = (\mu - 1)\left\{\frac{1}{r} - \frac{1}{s}\right\}.$$

From which we can say

$$\frac{1}{f} = (\mu - 1)\left\{\frac{1}{r} - \frac{1}{s}\right\}.$$

When applying the formula the correct signs must be allotted to r and s. They will have unlike signs in the case of a biconcave or biconvex lens, and like signs in the case of a meniscus or a concavo-convex lens.

Chapter Twelve

THE EYE AND OPTICAL INSTRUMENTS

REFERENCE was made in an earlier chapter to the photographic camera, which in essentials consists of a convex lens mounted in the front of a light-tight box, and giving a real inverted image on the farther side of the box where the sensitised plate is placed. The distance at which a screen should be situated in order to produce a well-defined image depends on the distance of the object from the lens. Hence a camera is made so that its depth can be altered at will, the process being known as "focussing". In order to overcome defects due to chromatic aberration (see p. 92) the lens of a camera is usually compound and is also corrected in other ways.

The human eye in many respects is similar to a camera, since it too consists of a combination of lenses which produce a real image upon a surface sensitive to light. A horizontal section is shown in diagrammatic form in Fig. 104. The following points should be noted:

The eyeball is roughly spherical, and about an inch in diameter. The front projects slightly with an increased curvature. There are three coatings:

(a) *The sclerotic*, a hard substance, the white of the eye, whose function is protective. It is modified in front to a clear transparent covering, the cornea, through which light passes into the eye.

(b) *The choroid*, which lies under the sclerotic. It consists largely of blood vessels and its function is to nourish the eye.

(c) *The retina*, which is the sensitive layer. It covers most of the interior but is absent from the front. A quantity of black pigment is present, which absorbs light and prevents glare. (For the same purpose a camera is painted a dull black inside.) The retina consists mainly of nerve cells.

The interior of the eye is divided into a shallow front portion and a nearly spherical chamber at the back by a ring of ciliary muscles, which suspend the crystalline lens, a transparent lens-shaped body, with its posterior surface more curved than the anterior. The choroid layer in front of the

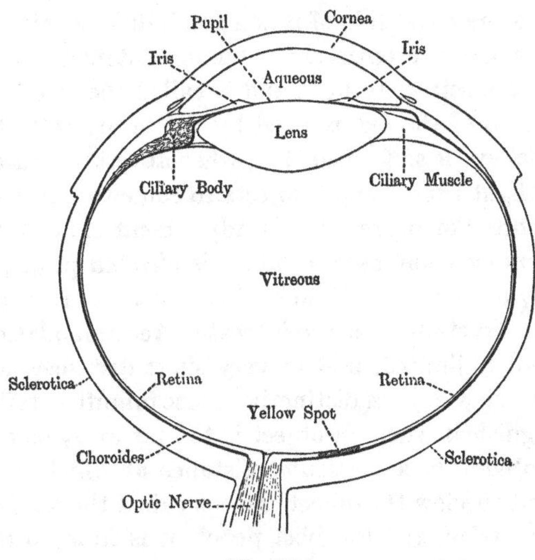

Fig. 104

lens has a circular aperture, the pupil, through which the black of the retina is visible. The aperture varies in size according to the amount of light entering the eye, and the choroid ring surrounding it, which imparts the characteristic colour to the eye, is called the iris. Between the crystalline lens and the cornea the space is filled with a dilute salt solution called the watery humour: behind the lens is a jelly-like substance, the vitreous humour.

The aqueous humour, lens and vitreous humour constitute the optical system of the eye. An image is formed on the retina, and opposite the lens is a small circular area coloured yellow, where the nerves are most sensitive. When we are looking at a small object—and it is a matter of common experience that although we can see much at a time we can only look at one point at once—the image of the point looked at comes on the yellow spot. Close to the yellow spot but slightly to the nasal side of it is a small disk devoid of nerve endings, where the optic nerve bundle enters the eye. This spot is not sensitive to light and is called the blind spot.

The depth of the eye is fixed by the horny sclerotic, and the normal eye is so formed that, when at rest, it causes plane waves of light from distant objects to come to a focus on the yellow spot. For nearer objects adjustment is necessary and this accommodation, as it is called, is effected principally by the ciliary muscles, which cause the curvature of the anterior face of the crystalline lens to increase. Accommodation however is not unlimited, and at very short distances it is not possible to view objects distinctly. Since minute detail cannot be distinguished when an object is too far away or too near, there is obviously a particular distance at which it is most convenient to view the object. This is called the *least distance of distinct vision*, and for most people it is from 10 to 12 in. from the eye. When looking at a book the reader automatically holds it in this position, and for purposes of calculation 10 in. or 25 cm. is taken.

From a consideration of the image formed by an ordinary convex lens it appears as though the image on the retina must be an inverted one. This is so, and the reinversion is effected in the sorting out by the brain of the nerve impulses it receives. An interesting illustration of this is based on the fact that if an object is held within the focal length of a lens the virtual image formed is upright. When viewing near objects the eye behaves as if its mean focal length were about 0·91 in.

A large pin-hole is made in a piece of thin card which is held about an inch from one eye, and a bright object such as a cloud is viewed through the hole. A pin is then held, with its head uppermost, between the hole and the eye. With a little adjustment it is possible to see the shadowy virtual image, and this image appears inverted.

The existence of the blind spot may be shown in the following way. A small black circle and a black cross are drawn on a piece of white card, about 10 cm. apart, and the card is then held about a foot away from the right eye, the line joining the marks being horizontal with the cross to the left.

The left eye is closed, and the paper moved to and from the eye, which gazes steadily at the cross. For one position of the paper the circle disappears. It may be necessary in certain cases to tilt the paper slightly to produce the effect. From the method used, it is obvious that the blind spot is situated between the yellow spot and the nose.

When a lighted taper is suddenly held in front of the eye of someone who is gazing steadily at a distant object, contraction of the iris, thus lessening the aperture, is apparent. When the eye is again steady it is possible to see three reflected images of the taper, one from the cornea, one from the front surface of the lens and one from the back of the lens. If the person then looks at a near object the image from the front of the lens is seen to decrease, showing that the curvature of the surface has increased, while the other images do not change: thus illustrating the process of accommodation.

Each eye receives light and an image is formed in each. So long as these images are formed on corresponding areas of the retina, the brain, in addition to the inverting process already mentioned, combines the two images, thus giving the impression of one. If however the images are not formed on corresponding areas the result is that we "see double". To show this, stand before a window and hold a pencil, about a foot away, between the eye and one of the uprights of the window. Look at the upright and two images of the pencil are seen. Then look at the pencil and details of the window are seen in duplicate.

When directed towards a solid object such as a tree trunk, the views obtained by either eye differ slightly. Each eye sees "round the corner", as it were, to a different extent. The result is the effect of solidity known as stereoscopic

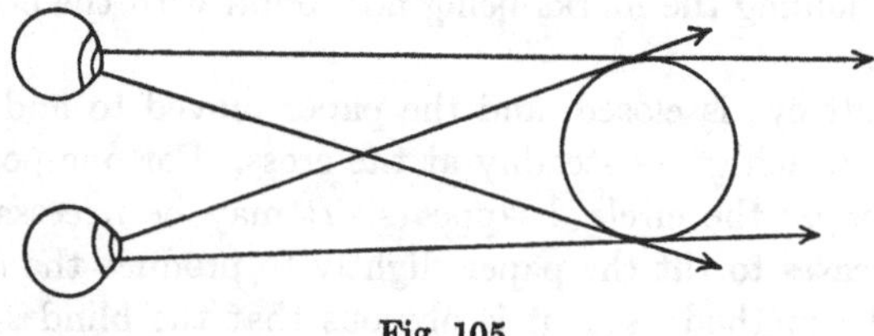

Fig. 105

vision, and will be understood from the diagram. The slightly different views obtained enable us to judge the relative distance from us at which objects are situated. A clear proof of this is obtained if a person covers up one eye, and then walks quickly up to a table on the extreme edge of which placed a threepenny-piece, and tries with a downward sweep of the arm to knock the coin off with his extended forefinger. The surprising difficulty encountered is due to the loss of a means of judging distance.

The instrument that at one time was very popular, known as the stereoscope, relies on the principle of binocular vision for the production of the well-known "solidity" effects in photographs. Two photographs are taken simultaneously by means of a special camera with two lenses, the horizontal distance between which is equal to the distance between the eyes. The views thus obtained correspond to what is seen by either eye. These are superimposed, using specially constructed half-lenses mounted in front of the pictures. The diagram will make clear the method of operation.

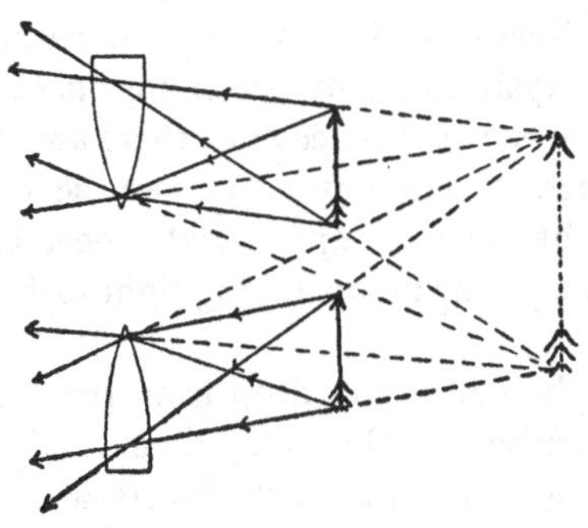

Fig. 106

Another phenomenon worthy of mention is persistence of vision. An impression formed by the retina lasts for about 1/10 of a second. Thus if more than ten impressions are viewed per second the eye is unable to follow and the results are blurred. Newton's colour whirler depends on this for its operation, as does the well-known toy, the thaumatrope, whereby two separate pictures such as a bird and a cage, drawn on the two sides of a card, can be combined by rapidly rotating the card about its longest axis.

When the stimulation on the retina is very intense, fatigue is caused. This accounts for the production of the green streaks which result from glancing at the sun. The experiment may be made of looking for 20 seconds at a brilliantly illuminated blue star painted on cardboard. On transferring the gaze to a white wall not too brightly lit, a yellow shadowy star is seen, *i.e.* the same pattern in its complementary colour.

The nerves engaged in registering blue become fatigued over the area covered by the image, and thus blue is missing from the sensations when the eyes look at the white wall. The other colours of this type are produced in a similar manner.

DEFECTS OF VISION

Apart from diseases of the eye such as cataract, which is caused by the crystalline lens becoming opaque, and which are outside the scope of this account, there are three principal structural defects which nowadays can be corrected with great precision by means of appropriate spectacles. These are known as short sight or myopia, long sight or hypermetropia, and astigmatism.

Myopia. If the eye is too deep from front to back, or if the posterior surface of the crystalline lens is too greatly curved, plane waves from distant objects will be brought to a focus somewhere in the vitreous humour, and only when the light is coming from near objects will the image recede to the retina. Thus blurred images will be formed except for very near objects. This can be corrected by imprinting a divergence on all light reaching the eye, *i.e.* by using spectacles of concave lenses. Such lenses are known to opticians as negative lenses, a -3 lens for instance imprinting a divergence of 3 diopters.

The strength of lens required will depend on how far away an object can be placed so as still to be seen distinctly. Suppose for instance that anything beyond 50 cm. is blurred for a particular person. This distance is called the "far point", and as for this position the eye is unaccommodated it corresponds to infinity for the normal eye. Thus a lens should be used which makes light actually from infinity appear to come from the far point, 50 cm.

A consideration of Fig. 107 shows that in order to do this the lens must have a focal length of 50 cm., *i.e.* equal to the distance of the far point, and be diverging. The result could be obtained by calculation as follows. Only light with an initial divergence of at least -2 diopters can be focussed, therefore a -2 lens must be placed in front of the eye.

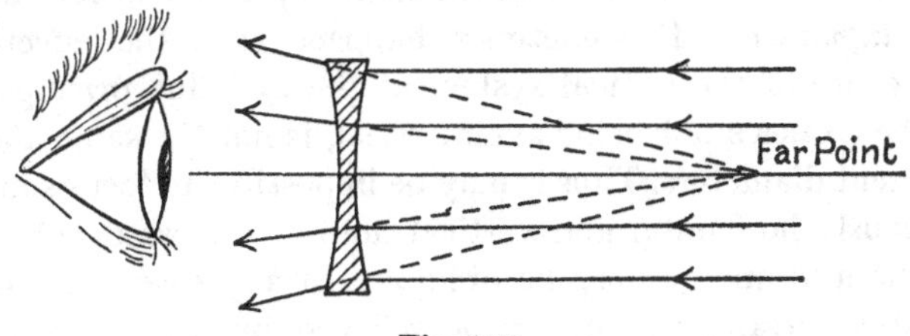

Fig. 107

Hypermetropia. The converse of the above, caused by the eye's being too shallow or by insufficient curvature of the posterior surface of the crystalline lens, can be corrected by imprinting an initial convergence using a convex lens. Old people frequently suffer from this defect as the eye begins to lose its power of accommodation, and thus they need spectacles for reading, although for ordinary purposes their sight may be perfect.

In order to illustrate the method used for finding how powerful a lens is needed, suppose that the patient cannot see clearly anything within 75 cm. of the eye. The convenient least distance of direct vision is 25 cm. (or 10 in.). Light from the 75 cm. point has a curvature of $-100/75$ diopters. Light from the 25 cm. mark has a curvature of -4 diopters. It is necessary therefore to alter this -4 and make it into $-100/75$.

Thus, if F is the required power,

$$-4 + F = -1\tfrac{1}{3}.$$

Whence

$$F = 2\tfrac{2}{3},$$

i.e. F is a converging lens whose power is $2\frac{2}{3}$, or, expressed as a focal length, is $100/2\frac{2}{3}$ cm. = 300/8 or $37\frac{1}{2}$ cm. If it is desired the ordinary lens formula $V = U + F$ could have been used. V is the curvature required for the image, $-100/75$, U is the curvature from the object, -4, and F the focal power of the lens. Thus the same expression is obtained.

Astigmatism. It sometimes happens that the effective curvature of the optical system of the eye, due principally to the cornea's not being symmetrical, is not the same along different diameters. Thus it may be impossible to focus simultaneously horizontal and vertical lines of a crossed pattern, or the asymmetry may be oblique. In any case there is a constant strain, which can be relieved by using spectacle lenses which have cylindrical surfaces, and so adding or subtracting partial curvatures as to cancel out the defect.

OPTICAL INSTRUMENTS

One of the oldest and most common uses of a lens is as a magnifying glass. Thus the first instrument to be dealt with is a glass of this type, also known as the Simple Microscope.

The size that an object appears to be depends on the angle it subtends at the eye. Thus the nearer it becomes the larger it appears, but for distances less than 10 in. or 25 cm. the eye is unable to accommodate itself. The use of the lens is such that an object can be brought close to the eye and the virtual image formed by the lens is thrown back to a distance of 10 in. without altering the subtended angle.

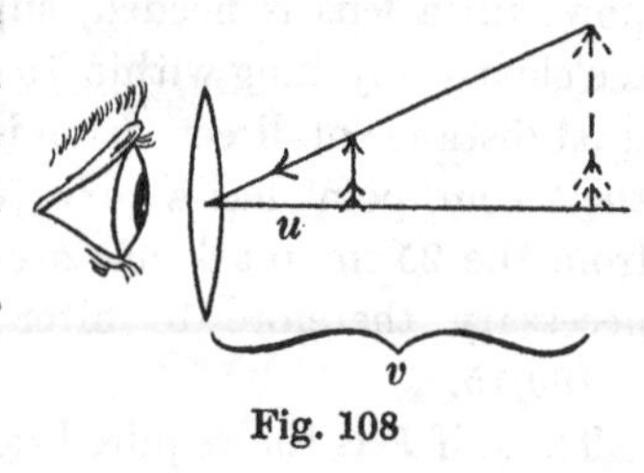

Fig. 108

This can be shown most simply in a diagram if we employ rays.

$$\text{Now the magnification} = \frac{\text{distance of image}}{\text{distance of object}} = \frac{v}{u}.$$

The ordinary lens formula $V = U + F$ can be written

$$\frac{1}{v} = \frac{1}{u} + \frac{1}{f}.$$

If we multiply through by v we obtain

$$1 = \frac{v}{u} + \frac{v}{f}$$

and $$\frac{v}{u} = 1 - \frac{v}{f},$$

i.e. $$M = \frac{f - v}{f}.$$

Now the distance v is 10 in. or 25 cm., and the light from it is diverging and therefore the sign for v is negative. Therefore

$$M = \frac{f + 10}{f} \text{ in inches or } \frac{f + 25}{f} \text{ in cm.}$$

Another expression may be obtained as follows:

Magnification $$\frac{v}{u} = \frac{U}{V}.$$

Now $$V = U + F,$$

divide by V, and $$1 = \frac{U}{V} + \frac{F}{V}.$$

Therefore $$\frac{U}{V} = 1 - \frac{F}{V}.$$

V is -4 (since light diverges from 25 cm. away). Therefore

$$\text{Magnification} = 1 + \frac{F}{4}.$$

It should be noted that these proofs assume that the eye can be regarded as touching the lens. The greatest magnification will occur when the light emerges from the lens almost parallel, *i.e.* when the object is almost at the principal focus.

The astronomical telescope.

This instrument, which was first investigated by Kepler, consists of two convex lenses, an object glass of long focal length which gives a real inverted image at its principal focus, and an eyepiece lens which magnifies the image in the manner already described. The diagram indicates how light from the top X of the image converges to the real image x and is thrown back to the position of distinct vision by the eyepiece. The objective in practice is always an achromatic combination and the eyepiece is compound to correct the various defects of a single lens.

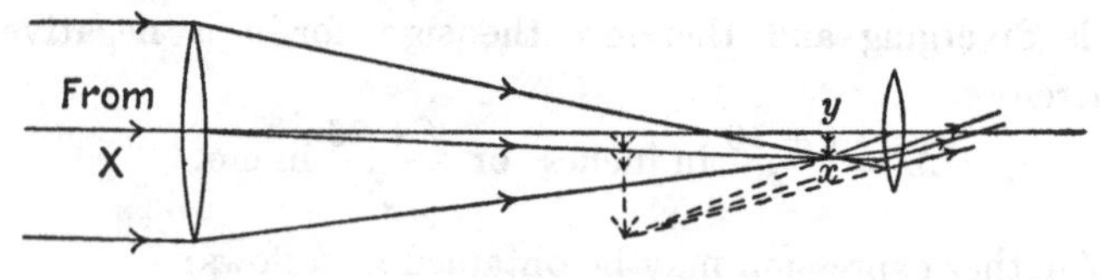

Fig. 109

An approximate value for the magnification may be obtained as follows. The angles subtended by the real image and the virtual image at the eye are the same, and the real image may be taken to be at the principal focus of each lens. Call the focal lengths f_0 and f_e respectively for objective and eyepiece. Suppose that the object is at a very large distance L. Then the angle it subtends at the objective $= XY/L$ in circular measure, and the angle at the eye may also be taken as XY/L.

But the image subtends an angle of $\frac{xy}{f_e}$. Thus

$$\text{Magnification} = \frac{\frac{xy}{f_e}}{\frac{XY}{L}}.$$

If we consider the objective alone, then

$$\frac{\text{Size of object}}{\text{Size of image}} = \frac{XY}{xy} = \frac{L}{f_0}.$$

Substituting in the above expression,

$$\text{Magnification} = \frac{f_0}{f_e} = \frac{\text{focal length of objective}}{\text{focal length of eyepiece}}.$$

The Galilean telescope.

Galileo constructed independently the telescope known by his name. Instead of a second convex lens the eyepiece consists of a concave lens and the distance between the two can

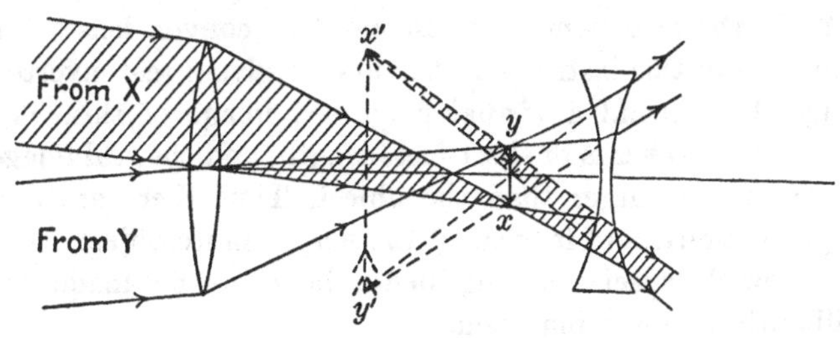

Fig. 110

be much shorter. Hence opera and field glasses were constructed after this pattern until the invention of the prismatic glass made it possible to use the principle of the astronomical telescope, with its greater magnification, in a confined space. The course of rays through a Galilean glass is shown above. As will be seen the virtual image is erect, and this of course is essential in a glass for terrestrial use. The magnification in this case also is given by the ratio of the focal lengths of objective and eyepiece.

The prismatic field glass.

In this case the necessary distance between the lenses is obtained by internal reflection in prisms, which also reinvert the image. The simplified diagram indicates the passage of a parallel beam through such a system of prisms.

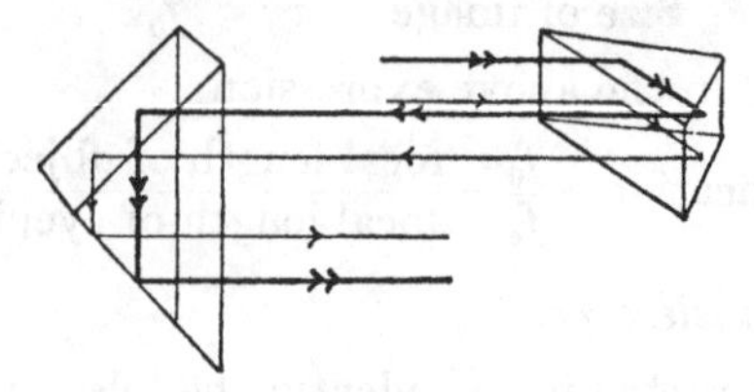

Fig. 111

The compound microscope.

As in the astronomical telescope, two convex lenses are employed in this instrument, the first to give a real inverted image, the second a virtual magnified image of this. The objective however is of short focal length, and the real image is very much larger than the object. Thus there are two magnifications: and in view of the very minute objects that are viewed, special arrangements have to be made for brilliantly illuminating them.

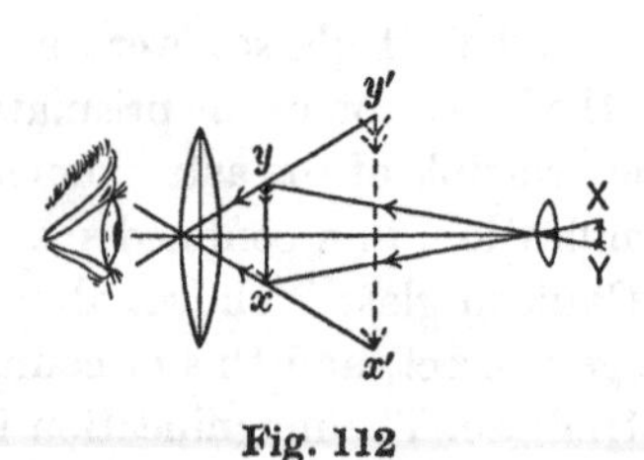

Fig. 112

Fig. 113

A simplified diagram is given in Fig. 112. In actual practice both the objective and eyepiece are corrected for chromatic and spherical aberration, and the complicated nature of a high power objective can be seen from the sectional view given in Fig. 113.

The magnifying power of such an arrangement may be computed as follows:

Call the tube-length, *i.e.* the distance between the objective and eyepiece, L, and the two focal lengths f_0 and f_e.

The greatest magnification will occur when the object is approximately at the focus of the objective and the real image at the focus of the eyepiece.

Then first magnification, due to objective,

$$M_0 = \frac{xy}{XY} = \frac{L - f_e}{f_0}.$$

And second magnification, due to eyepiece,

$$M_e = \frac{x'y'}{xy} = \frac{10\text{ in.}}{f_e}.$$

Thus magnification $M = M_0 \times M_e$

$$= \frac{10}{f_e} \times \frac{L - f_e}{f_0}.$$

This value is only approximate. It will be seen that the magnification is directly proportional to the length of the tube, and inversely to the focal lengths of objective and eyepiece.

In practice the focal length of the objective is made very small, the highest power in general use having a focal length of 1/12 in.

The reflecting telescope.

Newton's belief that it was impossible to construct lenses free from chromatic aberration led him to design and construct the telescope known by his name. Instead of an objective he used a large concave mirror, with a paraboloid surface to overcome spherical aberration. The reflected light was caught on a small plane mirror and turned through a right angle and the virtual image in the mirror was viewed through an eyepiece in the usual way. Other reflecting telescopes differed in the arrangement of the eyepiece. A diagram of the Newtonian reflector is given below.

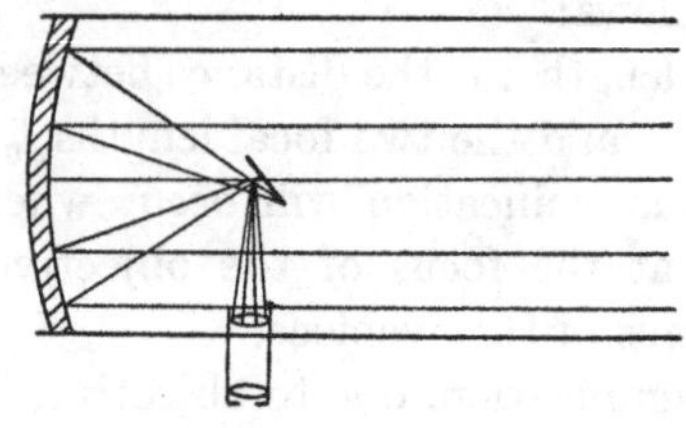

Fig. 114

The sextant.

In order to determine the latitude of a ship that is out of sight of land it is necessary to measure the angle made by the sun's rays with the horizon at noon. Instead of the sun's light that from certain stars may be employed, the *Nautical Almanac* giving tables whereby, from the observed elevation of the star at any particular time, the latitude can be calculated. The difficulty of measuring the angle from the moving deck of a ship is overcome by means of the sextant. The telescope T is directed towards the plane mirror E, the lower half only of which is silvered, and the horizon is viewed through the transparent portion. Cross-wires in the eyepiece help to fix the position of the image.

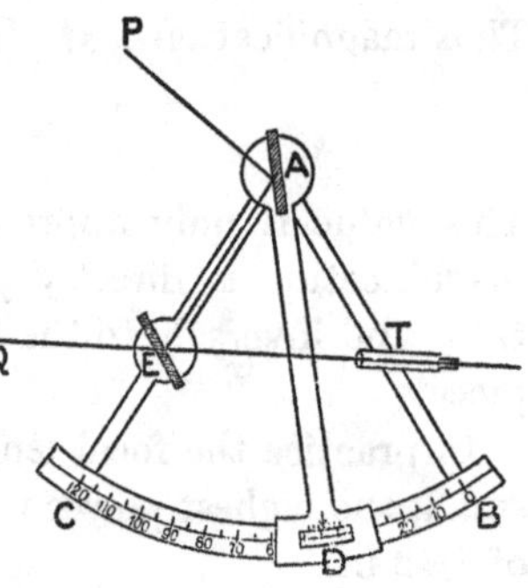

Fig. 115

The pivoted mirror A is turned by means of the arm D until the light from the horizon, after reflection at A and E, also enters the telescope, and when the two images coincide the position of D is noted on the scale.

The mirror A is now turned until light from the star or the sun after reflection forms an image coinciding with the horizon and the new position of D is read. Call the difference a degrees.

Since when a mirror is turned through any angle the reflected beam is turned through twice that angle, the difference between the light from the horizon and that from the sun must be $2a$ degrees.

By means of verniers and microscopes a can be read very accurately and so the angle of elevation of the sun is computed. Actually each half degree on the scale is labelled as a whole degree to save time.

The projection lantern.

The source of light in a lantern must be small and so an arc or else a bunched filament type of lamp is used. A powerful, slightly converging beam is obtained by means of two plano-convex lenses called the condensers, and after passing through

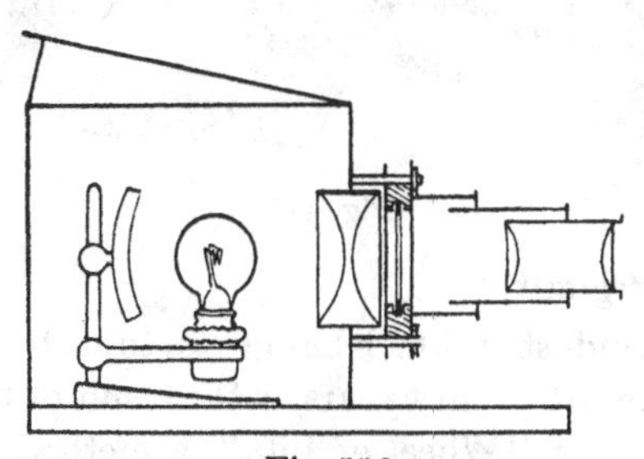

Fig. 116

the slide reaches the objective, a corrected system of lenses which forms a magnified real image on the screen. The arrangement is illustrated in the figure. The opaque-object projector, which has come into prominence recently, will be readily understood from Fig. 117. Great brilliance is obtained by the use of a mirror silvered on its front surface.

Fig 117

The cinematograph.

The cinema industry, which has grown to such enormous importance in the last twenty years, is the result of the invention of the Zoetrope or "Wheel of Life", a method of obtaining motion pictures, by an Englishman, W. G. Horner, in 1833.

The principle employed nowadays is as follows. By means of a special camera a series of instantaneous photographs is taken at intervals of about 1/20 of a second, on a long strip of sensitised film. This is developed to form a "negative" from which "positive" films are printed.

The projector has an optical system resembling that of

the ordinary lantern. The lamp is very powerful and the condensers form an intense narrow beam about 1 sq. in. in cross section which passes through an aperture known as the gate. The film passes in front of this gate with an intermittent motion, such that the different pictures are placed in the beam in turn, held there for a short interval and then jerked away again, about twenty successive pictures passing each second. The gate is fitted with a special shutter which closes while the film is moving and opens when each picture is in position.

The successive images on the screen follow one another with such rapidity that they appear to blend and so an appearance of animation is produced. This form of art lends itself to many novel and spectacular effects, and the technique of photographic production has reached a very high level, especially in America.

If an exceedingly rapid lens is employed, it is possible to photograph at many times the speed at which the pictures are to be shown, and the curious slow-motion effect is produced. On the other hand, by taking a few exposures at comparatively long intervals of a slowly altering subject, such as an opening flower-bud, it is possible to show on the screen during a few seconds the complete stages of a process that may actually require many hours. By these means the cinematograph lends itself to valuable analyses of motions which are either too rapid or too slow to be followed by the unaided eye.

The intense beam employed for projection carries a great many heat waves as well as those of light. Should the movement of the gate shutter and the film be interrupted the heat is sufficient to ignite the highly inflammable celluloid films. This has caused many fires in the past, and great efforts have been made to produce a non-brittle, non-inflammable substance for the film.

Chapter Thirteen

SOURCES OF LIGHT

PRIMITIVE peoples have always feared the darkness, and not without reason, for under the cover of night their animal foes could approach them unseen. When men discovered fire, some of the terrors were abated. In the friendly glare they were immune from attack; and later it was found that fats and oils would burn, so that it was possible to construct primitive lamps by floating a piece of wood or fibre in oil contained in a hollow vessel. The oils used were obtained from fish and animals, and the use of vegetable oils is as recent as the eighteenth century. Mineral oils are a still later adaptation and the paraffin lamp as we know it is an invention of the nineteenth century.

It was found moreover in early times that bundles of dry fibre dipped in fat made excellent torches. When a rush was dipped into molten tallow, which was then allowed to solidify, the rushlight, a primitive form of candle used by the Elizabethans, was produced. The substitution of a cotton wick came later, but even so the candles guttered sadly as the used wick drooped, and constant "snuffing" was necessary. The modern candle is so constructed that the wick curls over and burns away, and so snuffing is not required.

As Faraday points out in his book on the candle, its flame, as well as that of an oil lamp, is a gas flame, the function of the wick being to vaporise the oil or fat. Coal gas was produced at the beginning of the nineteenth century and was a great stride forward, although we should grumble nowadays were we forced to use the evil-smelling, dirty flames that then represented the last word in artificial lighting. The bunsen burner with its clean hot flame is a comparatively recent invention, and was followed by the use of mantles, invented by Otto von Welsbach. These consist of a network of ramie fibres impregnated with certain lime-like substances, chiefly thoria and yttria. When ready for use the fibre is burnt away and the earthy matter, on being raised to a high temperature by the bunsen flame, emits a brilliant white glow.

Davy was the first man to produce an electric arc of any size between carbon pencils, but arc lighting was not practicable until after the invention of the dynamo, due to Faraday's researches. The arc is produced by a powerful current jumping across a small air gap between two carbon rods, and the source of light is principally the tip of the positive rod, which becomes white hot. More recently flame arcs have been made, the carbons being impregnated with various metallic salts, and in these it is the glowing vapour between the rods which produces the light.

The electric glow-lamp, in which an enclosed filament is heated by the passage of a current, was first used by Grove in 1847, his lamps having platinum filaments. The carbon filament lamp was the first lamp used commercially, and then for a short time osmium filaments were tried, but the metal was too expensive. Tantalum filaments were in vogue for a time until 1911, when a method of filament construction using tungsten was perfected. Improvements in metallurgy and construction resulted in more durable lamps, and nowadays lamps with a filament of this metal, enclosed in an inert gas, are in universal use. A discussion of the merits of such lamps is given later.

At first sight the improvements are enormous. But, as we shall see, even to-day our methods of producing light involve terrible wastage. In all these cases the production of light is due to incandescence, that is to say, something is made so hot that it begins to emit waves of visible light. During the heating-up process it is emitting invisible heat waves and it continues to emit these even while the light is given out.

The following quotation from one of Sir Oliver Lodge's lectures gives a vivid picture of what happens:

> We want a certain range of oscillation, between 700 and 400 billion vibrations per second: no other is useful to us, because no other has any effect on our retina....To get this we have to fall back upon atoms. We know how to make atoms vibrate: it is done by what is called "heating" the substance....We take a lump of matter, say a carbon filament or a piece of quick-lime, and by raising its temperature we impress upon its atoms higher and higher modes of vibration—not transmuting the lower into the higher, but superimposing the higher on the lower—until at length we get such rates of vibration as our retina is constructed for, and are satisfied. But how wasteful and indirect and empirical is the process. We want a small range of rapid vibrations, and we know no better method than to make the whole series leading up to them. It is as though, in order to sound some little shrill octave of pipes in an organ, we were obliged to depress every key and every pedal, and to blow a young hurricane.*

Even in the best of our methods the actual fraction of energy emitted as visible light is very small. If, following the definitions of mechanics, we call the efficiency of the lamp the fraction obtained by dividing the useful (visible) energy by the total energy used, then it is found that the most efficient electric lamp, the flame arc, has an efficiency of barely 1 per cent. In other words, the wastage is about 99/100 for even good sources of light.

The invention of the bolometer by Langley made it possible for him to explore the spectra of various sources and so discover the relative amount of energy in different regions. The following graphs, taken from Professor S. P. Thompson's *Light Visible and Invisible*, illustrate what he found. The sun gives out about five times as much heat as visible light, but actually the energy of the light is greater than that of the

* *Modern Views on Light*, Sir Oliver Lodge. Published Macmillan.

heat. The electric arc gives a graph in which the region of the visible spectrum shows a pitifully small fraction of the total energy. But the third graph, from the spectrum of the light of the fire-fly, shows that all its energy is emitted in the form of visible light, and moreover a greenish-yellow light

Langley's Curves for One Unit of Heat.

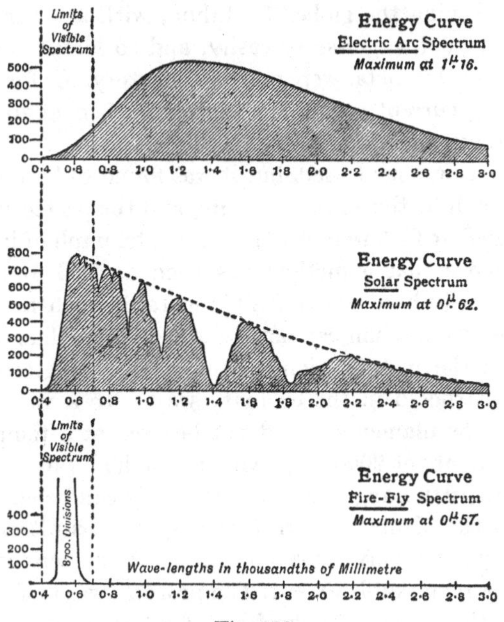

Fig. 118

which has the greatest possible effect on the eye. In other words, the insect has solved the problem of economical production of light, while man is still groping for a clue. But so far the insect has kept its secret, and we do not understand how it is done.

Langley's researches have shown that the hotter a source

becomes, the greater is the *proportion* of visible energy emitted. Hence the hotter we can make the filament of an electric lamp, the more efficient the lamp becomes; and research has been directed towards this end.

When a carbon filament lamp is overloaded with current it glows more brightly but the carbon vaporises and deposits on the glass, thus destroying the filament and at the same time blackening the globe. Tantalum, with a lower melting point, does not vaporise so easily, and so lamps containing filaments of this metal were more satisfactory. Unfortunately alternating current causes the tantalum filament to crystallise, thereby rendering it very fragile.

Tungsten was next used, but it was found to be impossible to draw it into the form of a wire, and the pasted filaments constructed at first were brittle. It is a triumph of ingenuity and patience that a method has been devised for treating this extremely hard and infusible metal in such a way that wires finer than a hair can now be drawn, using diamond dies.

Despite the metal's high melting point, tungsten lamps, when constructed in the ordinary way with a vacuum surrounding the filaments, could not be used at a temperature higher than about 2000° C., owing to the high rate of vaporisation. In 1913 Langmuir found that this can be checked by enclosing the filament in an inert gas, such as nitrogen, but then, owing to convection currents, the filament is cooled down to a remarkable extent. The problem was finally solved by using the filament in the form of a tightly wound spiral, when the convection effect is very largely checked, and what convection persists serves to carry away any vaporised metal and deposit it on a part of the globe where it will not matter. In this manner the modern gas-filled "half-watt" lamp has been produced, the filament withstanding a temperature of about 2500° C. The table shows relative efficiencies of electric lamps. The value of half watt per candle-power is only realised in the larger gas-filled lamps such as are used for projection and searchlights.

Type of lamp	Candle-power per watt	Watts per candle
Carbon	0·3	3·3
Tantalum	0·67	1·5
Tungsten	1·0	1·0
Gas-filled	2·0	0·5
Arc	1·5	0·67
Flame arc	5·8	0·17

Another type of electric lamp is the vapour-lamp, in which a gas is used for carrying the current and not a solid filament. Gases which have been rendered luminous, either by heating (as in the case of the sodium flame) or by an electric discharge, do not emit the whole range of the spectrum, but certain well-defined colours. Heat is not entirely absent but is very largely so. Mercury vapour and neon lamps are both common examples of this type, but although they are very efficient the light is somewhat distressing to the eyes.

The developments of the cathode-ray tube have led to the manufacture of a type of vapour-lamp which emits only ultra-violet radiation. When certain substances are placed in the path of such invisible waves they glow with brilliant colours. Here it appears as though we are on the threshold of what may lead to the production of cold light. The phenomenon is known as *fluorescence*.

In certain cases the substances continue to emit light even after the existing radiations cease. This property, which is made use of in the manufacture of luminous paint, is known as *phosphorescence*. The great drawback to the use of fluorescent or phosphorescent substances as sources of light is that the intensity of the emitted light is so feeble as to be of no practical value.

MEASUREMENT OF INTENSITY OF LIGHT

It is a matter of much importance to be able to compare the intensity of the light emitted by various sources. The original standard employed in this country was the Standard Candle, of spermaceti, six to the pound and burning 120 grains per hour. This has been superseded by various standard lamps, which however are still defined in terms of candle-power. One in use in this country is the Vernon-Harcourt 10 candle-power pentane lamp, but recently standard electric glow-lamps are becoming widely used.

For purposes of comparison the Inverse Square Law of Intensity is applied. The unknown lamp and the standard are caused to illuminate two specially prepared surfaces, and their distances are adjusted until the illuminations, as judged by the eye, are equal. The power of the lamp is proportional to the square of the distance at which the lamp is placed. Instruments used for comparison are known as photometers, and details of a few of the more important are given below.

The Bunsen or grease-spot photometer.

A sheet of white matt paper is rendered translucent at its middle point by a spot of grease. When the front is illu-

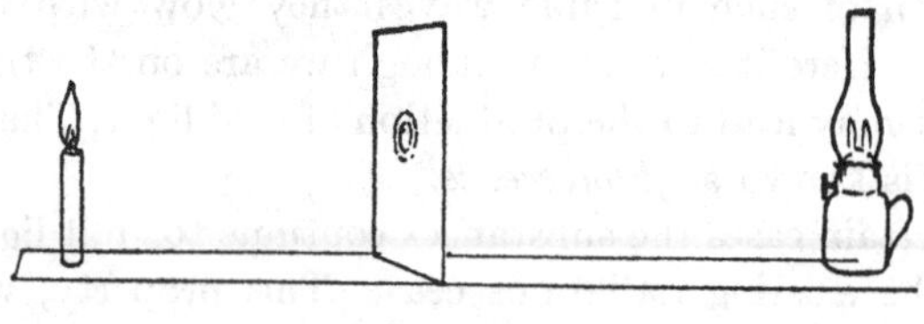

Fig. 119

minated the grease spot appears dark. When illuminated from behind the spot appears bright against a dark ground.

The standard lamp is placed at a fixed distance d in front of the paper. The unknown is placed behind and moved until the spot is least visible. Call its distance D. Then

$$\frac{\text{The power of lamp}}{\text{Standard lamp}} = \frac{D^2}{d^2}.$$

Joly's paraffin-wax photometer.

Two equal rectangular blocks of paraffin wax are separated by a sheet of tin-foil. The lamps are placed one on either side of the double block as shown, and are moved until the illumination of each block, as judged from the edge, is the same.

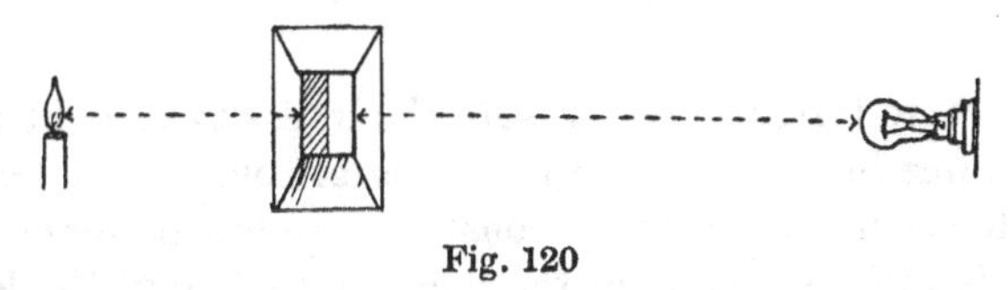

Fig. 120

The flicker photometer.

This is particularly useful when the two lamps for comparison differ in colour. A revolving disk of white card has two large sectors cut away, and is placed so that a sheet of

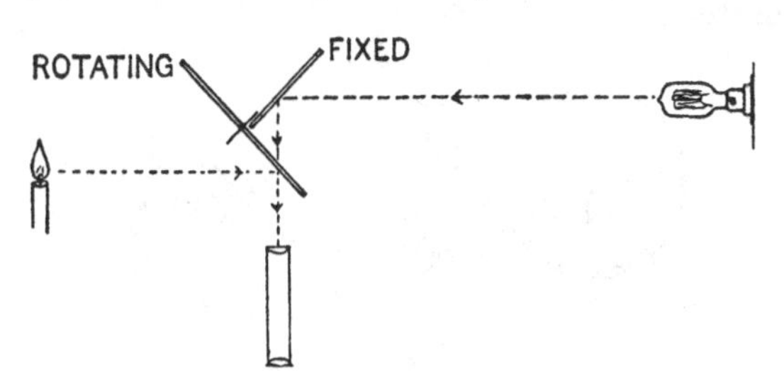

Fig. 121

similar card can be seen through one of the gaps. The disk and the other card are placed so as to make equal angles with the line of vision of the observer, and are illuminated respectively by the two lamps. When no flicker is seen as the disk revolves the two illuminations are taken as equal.

For descriptions of the precision photometer of Lummer and Brodhun, now in general use in industry and research, large text-books should be consulted.

In practice it is usual to employ the method of substitution in photometry, the standard being first balanced against a comparison lamp of no definite candle-power, and the distance of the standard read off. The lamp to be tested is then substituted for the standard and balanced. In this manner, errors due to asymmetry of the photometer head are eliminated.

All these instruments measure the intensity of light from the source in one direction only, and of course the intensity is different in different directions. A 50 candle-power electric lamp is rated in terms of the light emitted from the lower end, but naturally the cap end emits none: and so the average "spherical" candle-power is a different matter. Sphere photometers have been designed to measure this quantity, but a better measure of the strength of a source is the total light flux emitted.

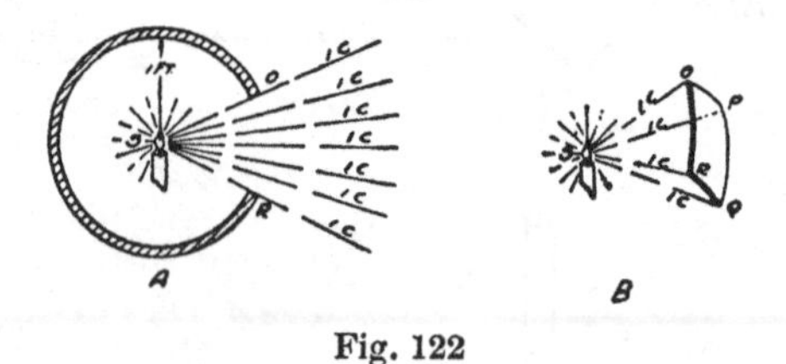

Fig. 122

The unit of light flux is the *lumen*. This is the light falling on each square foot of a spherical surface of radius 1 ft., around a standard candle. Since the area of such a sphere is 4π sq. ft., a standard candle emits 4π or 12·57 lumens.

Lumens generated by various illuminants.

Type of lamp	Watts	Total lumens	Efficiency in lumens per watt
Enclosed carbon arc	550	3100	5·6
Mercury arc	400	5340	13·3
Magnetic arc	532	10750	20·2
Carbon incandescent	50	174	3·5
Tungsten	50	500	10·0
Gas-filled	500	9000	18·0

Efficiency measured in this way is to be preferred to the candle-watt value given on p. 139.

INTENSITY OF ILLUMINATION

The intensity of illumination of a surface depends upon two factors, the power of the light and the distance away at which it is placed:

$$\text{Intensity} = \frac{\text{candle-power}}{(\text{distance})^2}.$$

This must not be confused with the *brightness* of a surface, which is dependent on the intensity and the degree to which the surface reflects light. The unit of intensity is the *foot-candle*, *i.e.* the illumination of a surface 1 ft. away from a standard candle. To satisfy this condition in theory the surface must be spherical. From the definition of the lumen already given it will be seen that

$$\text{Illumination in foot-candles} = \frac{\text{total lumens from source}}{\text{area to be lit}}.$$

This relation is of great importance in calculating the light sources to be used for various purposes.

Of recent years the importance of adequate illumination, both for industrial and home purposes, has come to be realised, and a considerable amount of research has been carried out to discover what is the most desirable value in different cases. A convenient form of instrument with which such measurements can be made is the foot-candle meter shown in Fig. 123. It consists, in essentials, of a trough, lit at one end by means of a small electric lamp, and covered by a screen containing a number of translucent spots which act as grease-spot photometers. When the illumination on the upper surface is within the range of the instrument, there will be a point on the under surface where the illumination is the same as that of the upper, and thus there will be one spot which merges into the background, those to the left being darker and those to the right lighter than the screen. The intensity corresponding to this spot is read off directly on the adjoining scale.

Observations have shown that the illumination under various conditions is as follows:

Full sunlight	7000 to 10,000	foot-candles
Mean day-lit interior in May	30 ,, 40	,,
Theatre stage	3·0 ,, 4·0	,,
Well-lit room	1·0 ,, 3·0	,,
Picture gallery	0·65 ,, 3·5	,,
Railway platform	0·05 ,, 0·5	,,
Full moonlight	1/100 ,, 1/60	,,

For comfort in reading the illumination should not be less than 5·0 foot-candles. Nowadays factory owners are required to see that the lighting of staircases and corridors is not less than the minimum safety value of 0·25 to 0·4 foot-candles. Fine work, to be carried out efficiently, requires at least 5·0 foot-candles, and for very fine work an illumination of from 10 to 20 foot-candles is desirable.

Fig. 123. The Foot-Candle Meter

Chapter Fourteen

INTERFERENCE, DIFFRACTION AND POLARISATION

In discussing the various theories as to the nature of light in the earlier part of this book, frequent mention has been made of interference which, with the allied phenomenon of diffraction, has been cited as one of the principal proofs of the wave-like nature of light. Although anything like a complete account of these would involve theory of a very advanced nature, their importance in the development of our modern ideas is such that even the beginner ought to know something about them. For this reason these few short notes are added, together with a brief reference to polarisation of light, which proved an unfathomable mystery until light came to be regarded as composed of transverse waves.

INTERFERENCE

Suppose there are two exactly similar sources sending out waves from adjacent positions, and suppose the waves are exactly in phase: in other words, a crest of one train is transmitted just at the same instant as a crest from the other, and so forth. Such an arrangement is realised in the two prongs of a vibrating tuning fork, which give out air waves in this manner. A diagram will help us to explore the results. Let X and Y be the sources, and let the crests be represented by firm lines and the troughs by dotted lines.

It will be seen that along a line at right angles to and bisecting the line XY, there are a series of points marked O, O, O, where crests or troughs from each source arrive simultaneously. At these points, therefore, the crest or trough will be exaggerated and reinforcement of the wave will occur.

At the points marked +, +, +, however, a crest from one source arrives at the same instant as a trough from the other and the two will cancel one another out. Thus along two lines on either side of the central reinforcement there will be destruction of the waves, and farther out again reinforcement will occur.

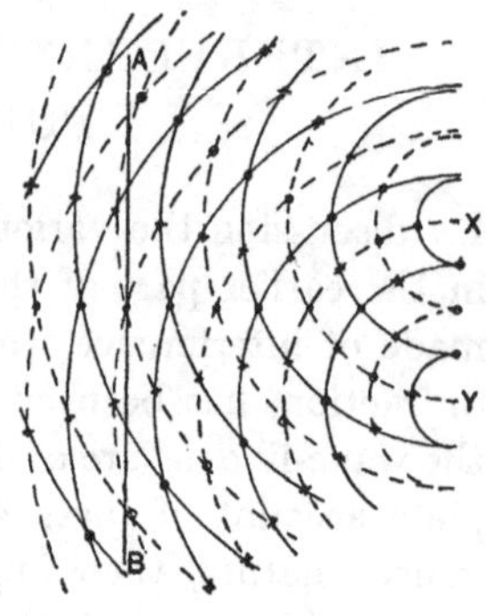

Fig. 124

Thus along a line *AB* there will be alternate regions, starting out on either side of the centre, of minimum and maximum disturbance. In the case of sound waves this means that there would be regions of alternate silence and sound. If a vibrating tuning fork is held upright, about 4 in. from one ear, and slowly rotated, the sound alternately dies away and swells again, as the regions of minimum and maximum disturbance pass the ear.

When dealing with light the regions of minimum disturbance should show an absence of light, that is, darkness. At first sight this seems absurd. If we stand two lamps side by side we do not get alternately bright illumination and darkness on the walls of the room. The apparent absurdity is due to the fact that for interference it is necessary to have two exactly equal sources *in phase*; and although our lamps may have the same candle-power there is little doubt that they would not be emitting troughs and crests simultaneously with the necessary degree of precision.

Once this has been grasped the following methods for providing interference of light will be understood. The first is a modification of Young's original method designed by Mr F. A. Meier, and can be performed in any room that has been partially darkened by drawing the curtains.

Young's parallel slits (modified).

The source of light is an electric 4-volt lamp with a straight-line filament. The screen S is an old photographic plate which has been fogged and developed, and on which two fine parallel cuts, 1/3 mm. apart, have been made with a safety

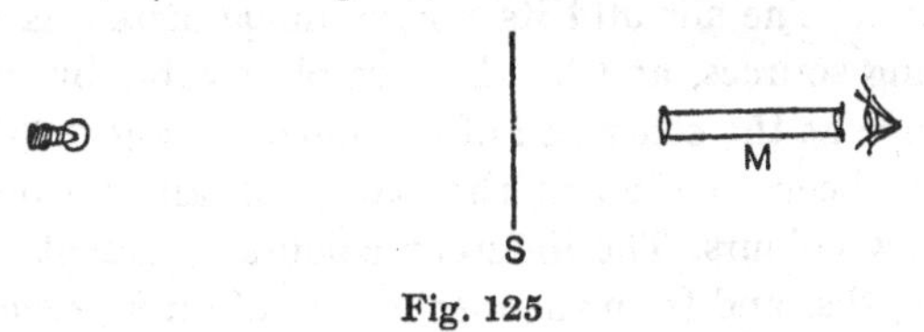

Fig. 125

razor blade. The distance between the slits and the lamp is 50 cm., with the glass side of the plate turned towards the source of light. A travelling microscope M is placed 15 to 20 cm. beyond the slits, and with very slight adjustments a marked series of vertical light and dark bands are observed. Some of the bands are coloured.*

Fresnel's biprism.

Fresnel used a prism with an angle just less than 180°. When this is placed in front of a narrow slit as shown in

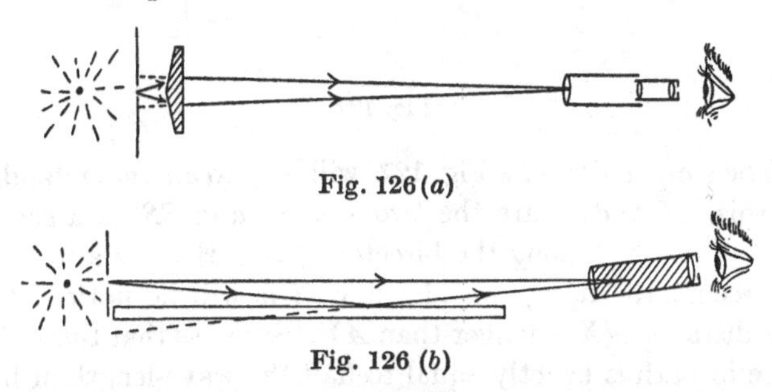

Fig. 126 (*a*)

Fig. 126 (*b*)

Fig. 126 *a*, two images are formed by refraction through the two sides, and as these are precisely similar very fine diffraction

* For further details see *School Science Review*, No. 30, Dec. 1926.

bands are obtained. When white light is used the dark portions are coloured, but if a monochromatic source of light is employed, such as a sodium flame, definitely black bands result. Dr Lloyd used a strip of plate glass 30 to 40 cm. long as a mirror, and at one end a slit, illuminated by a sodium flame, was placed. The slit and its mirror image acted as the two interfering sources, and bands were observable by means of a telescope at the other end of the mirror. (Fig. 126 *b*.)

As has been mentioned the bands obtained from white light show colours. The different colours represent different wave-lengths, and from the mechanism of interference waves of different lengths must interfere at different places. Thus if red waves interfere completely at one point the complementary colour will persist and a green band results; while if blue is destroyed by interference a complementary yellow is left.

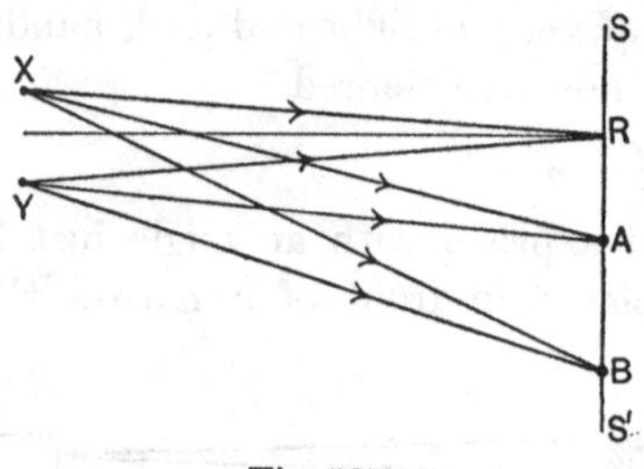

Fig. 127

The simple diagram, Fig. 127, will help to an understanding of this. X and Y are the two sources and SS' is a screen. At the point R along the bisector of XY there is always reinforcement. At a point A below R it will be noticed that the distance AX is longer than AY. Suppose that this difference in path is exactly equal to half the wave-length of blue light. Then blue will be destroyed by interference here, although other colours will be unaffected, and yellow is left. Farther down still a point B is reached such that XB is exactly half a red wave-length longer than YB. Then red is destroyed here and green results.

The most beautiful of all interference colours are those due to thin films, the best known being the colours of the soap bubble. Interference occurs between the light which is reflected from the front surface of the film and that which has penetrated the film and is reflected from the back surface. If the difference in path causes the second set of waves to be exactly out of phase with the former for any particular colour, that colour is destroyed and the complementary colour left. As the thickness of the film varies in different places, different colours are produced, and so a splash of dirty oil on a wet road can result in the magnificent ringed patterns so familiar in these days of motor transport.

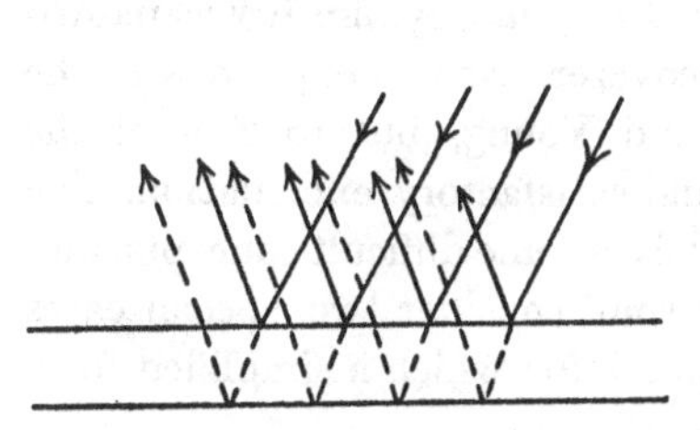

Fig. 128. Interference in a thin film. The path of the refracted light is indicated by dotted lines.

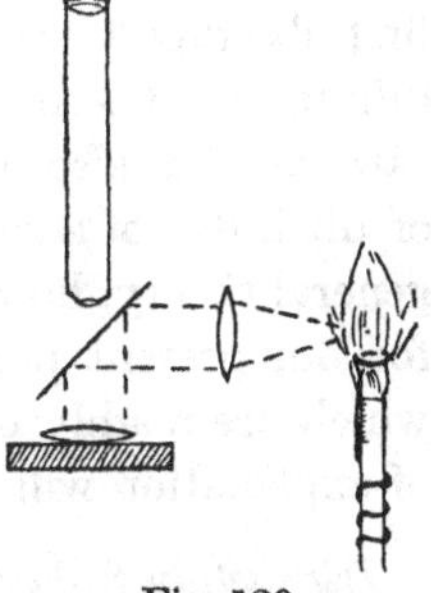

Fig. 129

Newton noticed and investigated an allied phenomenon that has been given the name of Newton's Rings. If a convex lens of rather long focus is held against a piece of plane glass a small dark spot is visible where the curved and flat surfaces are in contact. On holding the lens and glass up to the light and pressing them together, the spot is seen to grow larger and appears to consist of a series of concentric coloured circles.

If the two are illuminated by monochromatic light as shown in Fig. 129, the microscope reveals the presence of many scores of dark concentric rings. They are produced by interference between light reflected at the upper and lower surfaces of the air film, which gradually becomes thinner as the point of contact is approached, thus giving rise alternately to the conditions necessary for reinforcement and mutual destruction. The diagram indicates how the phase difference is brought about.

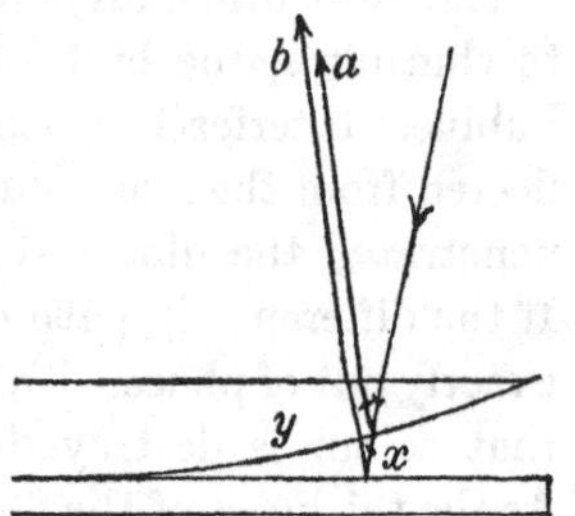

Fig. 130. *a* and *b* mutually interfere. At a point *y* there would be reinforcement.

DIFFRACTION

The bending of light round into a shadow, which we should expect from a consideration of waves, was first observed by Grimaldi in 1665, and by him it was named diffraction. The original discoverer tried to explain what he saw, as did also Newton and Young, but to Fresnel the credit is due of a complete and satisfactory explanation. The general theory, however, involves some difficult concepts, and for our present purpose we shall consider two special cases which are readily observed, and for which a simplified form of explanation will serve.

Diffraction from a narrow slit.

A good source of light for experiments on diffraction is obtained by enclosing a frosted electric lamp in a cocoa tin, in which a narrow slit has been cut. If the sides of the slit are parallel it is then necessary merely to view this slit through another, obtained by cutting a fogged photographic plate, in order to see the diffraction bands.

These consist of a bright central band bordered on either side by alternate bright and dark coloured bands, the intensity falling off rapidly on either side of the centre. If an adjustable slit is available for placing in front of the eye, it is possible, by making the slit narrower, to make the bands broader, and vice versa.

The origin of the bands will be understood in part from a consideration of the figure. For the sake of simplicity suppose the light is yellow monochromatic light. *AB* is the narrow slit and *EE′* represents the retina of the eye. The light reaching the slit, from the nature of its production, may be considered as consisting of plane waves. Then as the

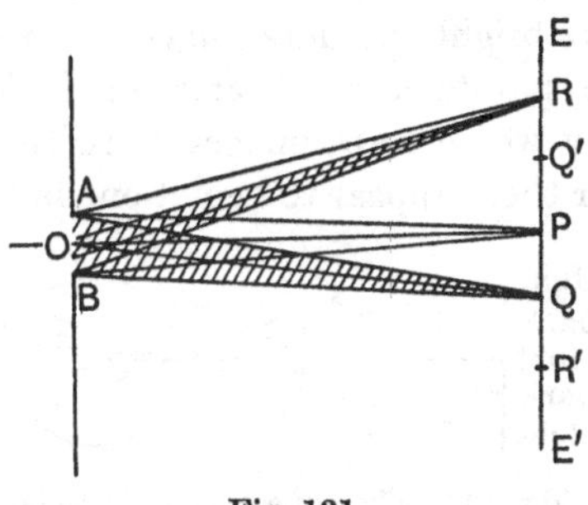

Fig. 131

slit is narrow, for a point *P* opposite its centre *O*, the distances *AP*, *OP*, *BP* will be all sufficiently alike for no interference to occur, and so a bright line occurs at *P* on the retina.

Next consider a point *Q* such that *AQ* is half a wave-length of yellow light longer than *OQ*, and *OQ* half a wave-length longer than *BQ*. Then for every element of one half of this beam, there is a corresponding element in the other half which is exactly out of step. Thus interference occurs at *Q* and darkness results. A corresponding point *Q′* exists on the other side of *P*.

For a point *R* still farther out (drawn above *P* in the figure for clearness), the paths *AR* and *BR* differ by $1\frac{1}{2}$ wave-lengths. Then if the beam is divided into three, there will be two mutually destructive portions and a third which will survive. Thus a bright band results which has roughly one-third the intensity of the central one. When the paths from the extremes of the slit differ by two wave-lengths, if divided into four the beam is seen to contain two pairs of quarter beams which mutually interfere, and so again produce darkness. On the outer edge there will be a region where one-fifth of the next beam (whose boundaries differ in length by $2\frac{1}{2}$ wave-lengths) will survive.

The alternate bright and dark images formed on the retina correspond to separate narrow beams of light from such directions that they would form images at these places. Hence to the observer there appear to be not one but many sources,

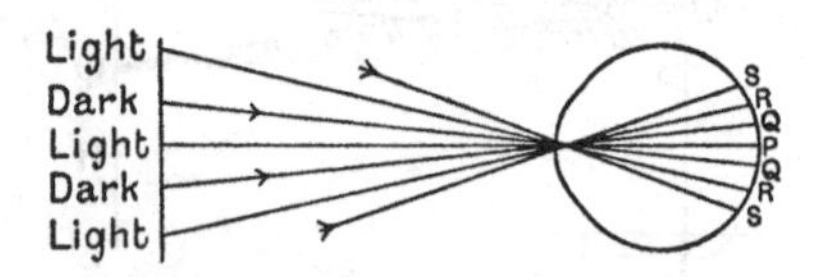

Fig. 132. The appearance of bands.

the central one being the most powerful. By narrowing down the slit the angles between the beams are increased and so the diffraction bands are made wider. The presence of colour in bands formed by white light will be readily understood from a consideration of interference.

Diffraction from a narrow obstacle.

The lamp described is set up at one end of the laboratory. A knitting-pin is placed about 6 ft. in front of it in a vertical position, and about 12 ft. beyond a telescope is placed so that the knitting-pin is in a direct line with the light.

A slight adjustment reveals that there are numerous vertical bright bands in the shadow of the pin, produced by interference between the light that has come past either edge of the pin.

The diffraction grating.

Professor Rowland in America invented a machine which is capable of ruling an enormous number of parallel lines to the inch on the surface of a speculum mirror. The result is known as a diffraction grating, and from it replicas may be constructed either as casts or by photography. The ordinary diffraction grating as used in laboratories consists of a transparent photographic reproduction mounted on plate glass. There are usually 14,500 lines to the inch.

Let the dotted line represent part of such a surface with light falling normally on it. From each slit a secondary wavelet commences, and those portions travelling straight on do not interfere, so that the grating is transparent to light along the normal.

Consider now light coming away from two adjacent slits *a* and *b* so that the combined wave-front *xb*, which is at right angles to the direction, will be a little distance *ax* farther from *a* than it is from *b*. If this distance is a whole wave-length of, say, yellow light there will be reinforcement, and this will be true also for light from every one of the slits. Thus yellow light will come away from the grating in a parallel beam in this direction.

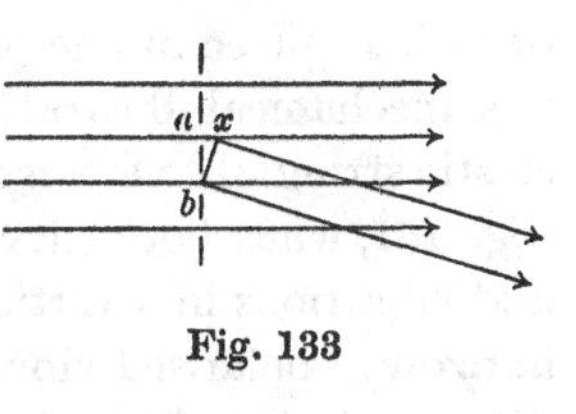

Fig. 133

For smaller angles there will be reinforcement of shorter waves, green, blue, indigo, violet, and for a large angle of the longer orange and red waves. Except in the particular direction in which there is reinforcement there will be complete interference for each colour: and so a spectrum is produced in which the longer waves are most deflected and the shorter waves least. The order of the colours is therefore the reverse of that produced by a prism.

Since the angle of deflection depends upon the wave-length, by measuring this angle and also the spacing of the lines it becomes a simple matter to calculate the wave-length of a particular colour.* This has been one of the most important uses of the grating. Moreover its dispersion is much larger than that of a prism and so by its means much has been learnt of the minute details of different spectra.

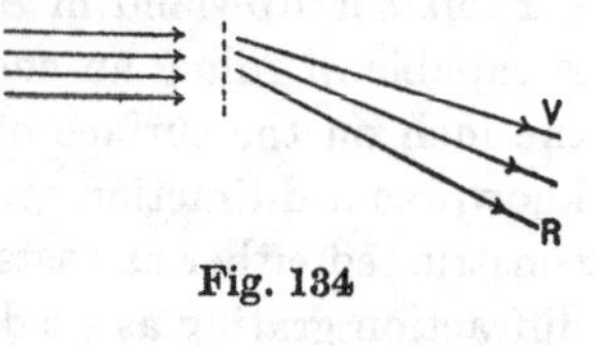

Fig. 134

POLARISATION

The waves of light are transverse, that is to say, the vibrations are at right angles to the direction of motion of the wave. There is a vast number of possible planes of vibration, up and down, side to side, obliquely and so forth. Vibrations of this sort are said to be unpolarised. If however the vibrations take place in one plane only they are plane polarised.

A mechanical illustration is afforded by the vibration of an elastic string. If we pass the string through a pair of slits as in Fig. 135, when both slits are vertical they will transmit polarised vibrations in a vertical plane. When both are horizontal, horizontal polarised vibrations can pass. If however the first slit, or polariser, is vertical and the second, or analyser, is horizontal, then no vibrations at all can pass beyond the second slit.

Many crystals possess the power of transmitting light at different velocities in different directions. Iceland spar shows this very clearly, the two velocity constants giving rise to two different refracted beams, so that a crystal placed on a printed page causes the type to appear in duplicate. This property is known as double refraction and was well known to Huyghens, who however was unable to explain it.

We now know that the two refracted beams are both plane

* For if λ is the wave-length, a the distance between two lines and θ the angle of deflection, then $\lambda = a \sin \theta$.

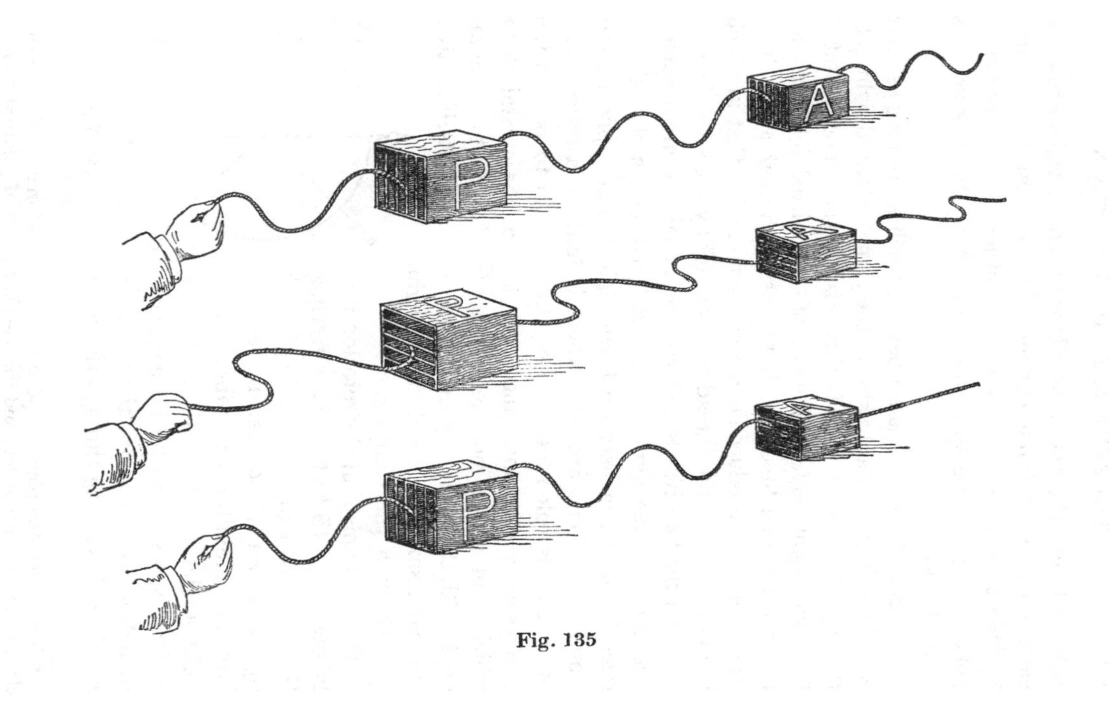

Fig. 135

polarised, and the planes of vibration are at right angles to each other. By suitable cutting of the crystal it is possible to deflect one beam aside, and so the other passes on as plane-polarised light. The specially prepared crystal is known as a Nicol prism.

If a second Nicol is placed beyond the first, so long as their corresponding axes are parallel the plane-polarised light gets through the second one. If this prism is then slowly rotated through a right angle the amount of transmitted light becomes less and less and finally no light passes through at all. This "dark field" is produced in a manner similar to that in which the vibrations of the thread were cut off by the second slit.

When the two Nicols, polariser and analyser, are in such a position that the dark field is produced, and a thin slip of mica is pushed in between them, light at once appears, and wonderful colours are seen where it falls on a screen. The mica itself is colourless. Its crystals happen to possess the two velocity constants, and if the axis of the crystal along which light passes most readily is placed so as to make an angle with the plane of polarisation, the plane is itself rotated through a small angle and so some light gets through the analyser.

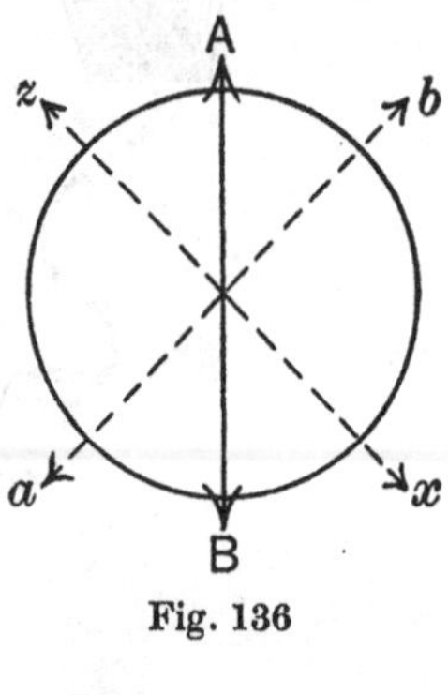

The mechanism of the rotation is as follows. Let AB be the plane in which the polarised light is vibrating.* This up and down motion can be resolved into two simultaneous vibrations at right angles to each other, along ab and xz. Now suppose the mica is placed with its axis of maximum rigidity along xz. Then, in passing through the mica,

Fig. 136

* This is *not* the plane of polarisation. Unfortunately the nomenclature is rather confusing, and the so-called plane of polarisation is at right angles to the vibrations.

the resolved part of the vibration along xz will get ahead of the resolved part along ab, so that on recombining the two the resultant vibration will be no longer along AB but will have rotated through some angle.

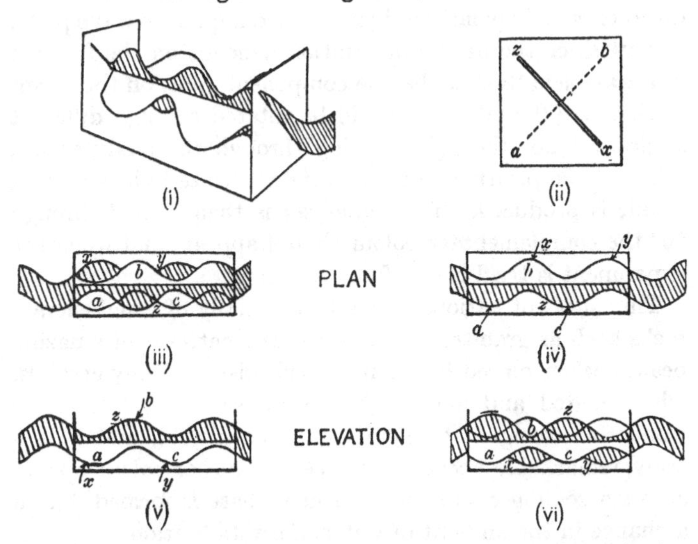

Fig. 137. Mechanical illustration of rotation of the plane of polarisation. (i) General view of apparatus. The shaded "sine curve" can slide lengthwise in slots. (ii) End view. The two curves are each at 45° to the vertical and represent the resolved parts of polarised waves, the plane of polarisation being either vertical or horizontal. (iii) Plan of apparatus and (v) side elevation in the position representing a vibration in a vertical plane only. (iv) Plan and (vi) elevation when the shaded component has been advanced half a wave-length. Note that the vibrations are now from side to side, *i.e.* the plane of polarisation has been rotated through 90°.

The drawings are of models designed by the author to illustrate this rotation. If the one vibration gets exactly a quarter of a wave-length in front of the other, then a line joining the crests goes in a spiral round the wave, which is said to be circularly polarised. When this is produced an equal amount will be transmitted for all positions of the analyser. Mica plates designed for use with polarised light are named according to the amount that one resolved part is advanced with respect to the other, and a plate that produces circularly polarised light is called a quarter-wave plate.

Now since the degree of rotation depends on the fraction of a wave-length that the one component gains on the other, obviously different colours will be rotated through different angles. Hence the light passing through the analyser will differ in composition from the original white light and so a colour is produced. If the analyser is then turned through 90° the complementary colour should appear; and when the experiment is tried this is found to occur.

Thin slices of various minerals, especially of conglomerate rocks such as granite, produce coloured patterns of amazing beauty when placed between crossed prisms. Many crystals, when melted and allowed to crystallise on a slide, show wonderful changes of colour in the dark field, and it is a truly remarkable spectacle to see a wave of colour spread across a cooling crystal when a new phase is formed due to a change in the amount of water of crystallisation.

Certain liquids too possess the power of rotating the plane of polarisation. The property is known in this case as optical activity, and a study of the activity exhibited by the various sugars has contributed much to the theory of organic chemistry. Optical activity is always associated with asymmetry of the molecule.

The foregoing sketchy notes do little more than define the terms interference, diffraction and polarisation. Through a study of these complex and fascinating phenomena modern theory has made great advances, and it is only in terms of waves that any attempt can be made to interpret what has been observed. Waves of light have been described as electromagnetic disturbances. We know nowadays that electricity has an atomic structure, its ultimate particles being known as electrons. The electrons are minute negative charges and repel one another, the direction of repulsion being known as a line of force. These lines may be pictured as passing out in all directions from an electron as in Fig. 138. A movement of electrons constitutes an electric current. Associated with a current there is always a magnetic force, in concentric circles round the current (see Fig. 139) and depending in direction on the direction of the current. Suppose now an electron commences to oscillate between two positions. As

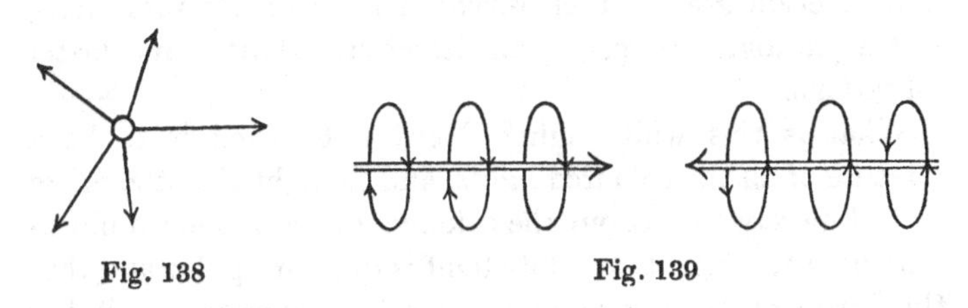

Fig. 138 Fig. 139

it moves down magnetic lines of force are set up in one direction. As it moves up their direction is reversed. Moreover as a result of the movement we can picture a kink appearing in each electric line of force. Thus a simultaneous electric and magnetic disturbance will result, and Maxwell has calculated that this must be transmitted through the ether at a velocity given by the ratio of the absolute electromagnetic and electro-static units. Experimental measurement shows that this ratio is about 300,000,000 metres per second. The resemblance between this and the measured velocity of light is too close to be a mere coincidence; and when Hertz succeeded in producing definite electro-magnetic waves, which are propagated with the speed of light, the views of Maxwell were confirmed in a striking manner.

The number of vibrations made by the electrons per second to produce visible light must be between 400 and 700 billions!

We cannot visualise what this represents. Sir Oliver Lodge has illustrated it in this manner: "The number...may be represented by the number of vibrations executed by a tuning fork sounding a note two octaves above middle C, continuing for 12,000 years".*

As we gradually heat up a substance such as a piece of platinum, the vibrations of the electrons become faster and faster. At first vibrations of long wave-length are given off, and are recognisable as radiant heat. After a time the metal becomes a dull red, then bright red and finally white hot. A diffraction grating then reveals all the visible radiations, and a photographic plate can detect still shorter and faster vibrations.

What is this white light? Newton believed it to be a mixture of all the coloured lights, and thought that the prism served merely to sort out the colours. Certain modern physicists hold the view that white light is quite irregular and that the prism or grating manufactures the colours; and it has been shown by theoretical analysis that this is a reasonable possibility. Let us confess with due humility that we cannot yet give a final answer. As our knowledge increases new problems present themselves and new horizons are seen; but with every advance we come to appreciate more and more the essential unity of nature, and as we find solutions to our problems we catch glimpses, however fleeting, of the majestic simplicity of the universe.

* *Modern Views on Electricity*, q.v.

Chapter Fifteen

SUGGESTIONS FOR CLASS EXPERIMENTS ILLUSTRATING SOME OF THE FUNDAMENTALS OF OPTICS

FOR thirty years the practical work connected with elementary light has been carried out with what have been termed "Pin and Parallax" methods. One great advantage has been cheapness: but the tedious nature of the operations has destroyed much of the value that sustained and lively interest could impart to the work.

Thanks to the skill of Mr F. A. Meier of Rugby in designing apparatus it is now possible for any school to obtain, at a cost which cannot be called oppressive, complete sets of apparatus whereby the pupils can carry out the elementary work using *real beams of light*. The help that the teacher receives from the added interest is something to which the writer can thankfully testify.

Fig. 140 shows a set of the apparatus.* The source of light is a 4-volt electric lamp, which is placed in a "condenser" consisting of a small tin in the end of which is fixed a bull's-eye lens. Apertures of different shapes, slits or circles, can be clipped in front of the condenser. Stands for clipping lenses and other apparatus are made from small bulldog clips into which a strip of rubber has been introduced. The screen is a strip of metal bent at right angles. Full directions for the construction of this apparatus, including cost, and for various experiments, are given in the *School Science Review*, No. 13, September 1922.

* As supplied by Messrs Philip Harris and Co., Ltd.

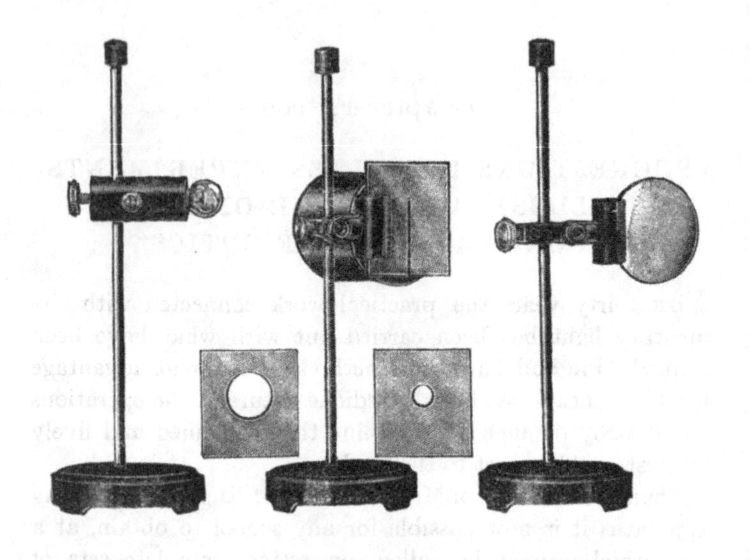

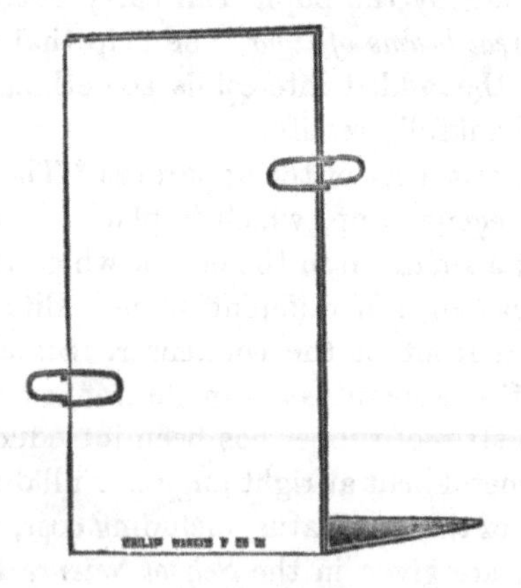

Fig. 140

For those experiments in which parallax provides the best method of adjustment, the luminous wire and black glass mirrors, which the writer first saw used by Mr Grace's classes at the Latymer School, Hammersmith, are to be recommended. The substitution of plain glass mirrors painted on the back, for the expensive black glass instruments, so far as the writer is aware, is original.

To set apparatus for production of a beam of light.

Adjust the lamp inside the condenser so that the metal filament is approximately focussed on a convex lens of 10 cm. focal length. Move the lens until a clear image of the slit is thrown on the screen 18 in. away. Place a drawing-board horizontally between the lens and screen. Lower the lens and a narrow regular beam is produced across the drawing-board. Do not stand the lens on the board, and work with one edge only of the beam.

EXPERIMENT 1

To illustrate the law of reflection of light at a plane surface.

Obtain a narrow parallel beam along the 0°–180° diameter of a divided circle. Place a strip of mirror* along the 90°–270° diameter with the polished surface facing the source of light. Observe that under these conditions the light is reflected back along its own path.

Now rotate the circle and mirror about the point of incidence, and obtain a number of values for the angles of incidence and reflection, *i.e.* between the incident and reflected paths and the normal.

Make a general statement covering all the cases.

* Instead of glass mirrors, with their inevitable errors, the writer has employed successfully etching plates of copper, electro-plated with chromium—in fact "stainless steel" mirrors, which are more durable than silver.

EXPERIMENT 2

Rotation of the mirror and the reflected beam.

Set up the apparatus as in the previous experiment and rotate the card so as to make an angle of incidence of 20°.

Now turn the mirror through an angle of 15° about the point of incidence. Note the angle through which the reflected beam is moved.

Repeat, using different angles, and state the general effect of rotating the mirror. Tabulate your results.

EXPERIMENT 3

Total internal reflection in a right-angled prism.

Obtain a beam and let it fall normally on one of the shorter faces of a right-angled prism. Observe the path of the reflected ray.

Now move the prism so that the incident beam is normal to the hypotenuse and again observe the reflected path.

Set two such prisms:

(*a*) as for use in a periscope;

(*b*) as for use in prism binoculars.

EXPERIMENT 4

Refraction of light through a semicircular block.

(*a*) Place a divided circle on the drawing-board and obtain a beam of light along the 0°–180° diameter. Place the block with its flat surface facing the light and along the 90°–270° diameter. Observe that under these conditions there is no refraction. Now rotate the drawing-board through any angle, keeping the point of incidence of the light on the glass at the centre of the circle. Observe the angle of incidence (*i.e.* the angle made by the original and the new path before reaching the glass) and the angle of refraction (*i.e.* the angle between the original and the new path in the glass).

Repeat, using different values for the angle of incidence.

(*b*) Now reverse the block so that refraction occurs only as the light leaves the glass. Slowly rotate the board until the emergent beam just grazes the surface of the glass. Measure the angle between the beam in the glass and the normal, when this occurs. This is called the critical angle.

Rotate the board the least bit more, and note that total internal reflection occurs.

Note. If the base of the block is painted white the path of light *in the block* is rendered visible.

EXPERIMENT 5

Refraction of light through a rectangular block (*or a rectangular tank*).

Rule a line on paper and let an edge of the beam lie along this line. Place the block across the line in such a way that the path of the light is not altered.

Rotate the board about the point of incidence and mark the paths of the incident and emergent beams. Rule a line along both faces of the block, remove the block and rule in the paths of the incident and emergent beams, and the beam in the block.

Measure the angles made with the normals by the two beams in air and both ends of the beam in the glass.

Repeat, using different angles and different thicknesses of glass, and tabulate your results.

EXPERIMENT 6

To measure the angle of a prism.

Obtain a beam along a pencil line on paper. Place the prism so that its angle is on the line, pointing towards the source of light (see Fig. 141).

Two reflected rays can be marked with pins.

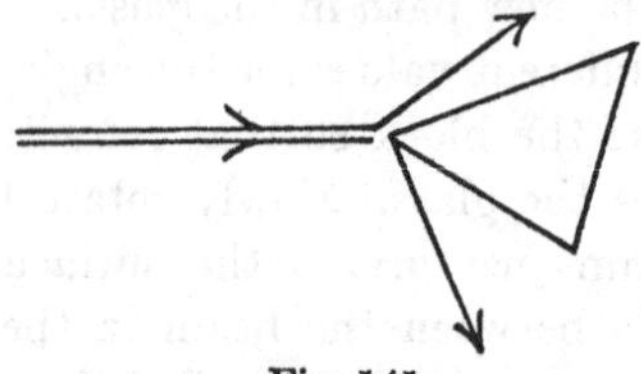

Fig. 141

The angle between these rays is twice the angle of the prism. Use the prism in at least two different positions.

Repeat for the other two angles of the prism.

EXPERIMENT 7

The refractive index of water.

Nearly fill a small crystallising dish with water. Draw on paper a circle the same size as the dish, and a diameter to the circle, and place the dish in the circle.

Place the board so that the beam strikes the dish at one end of the diameter. If the water contains a trace of fluorescein the refracted beam can be seen clearly, and a ruler can be laid on top of the dish with one edge over the beam. By means of a set square, the point at which the far end of the beam leaves the dish can be marked on the paper, and the incident beam should also be marked.

The dish is then removed, the beams ruled in, and the angles of incidence and refraction measured.

Then
$$\mu = \frac{\sin i}{\sin r}.$$

EXPERIMENT 8

Measurement of critical angle for water, using circular dish.

Use a 12-cm. crystallising dish full of water. Obtain a narrow circular beam, and by means of a short strip of mirror held on a thread throw this beam up to the surface of the water (see Fig. 142).

Adjust the thread until the patch of light on the table, due to internal reflection, just disappears.

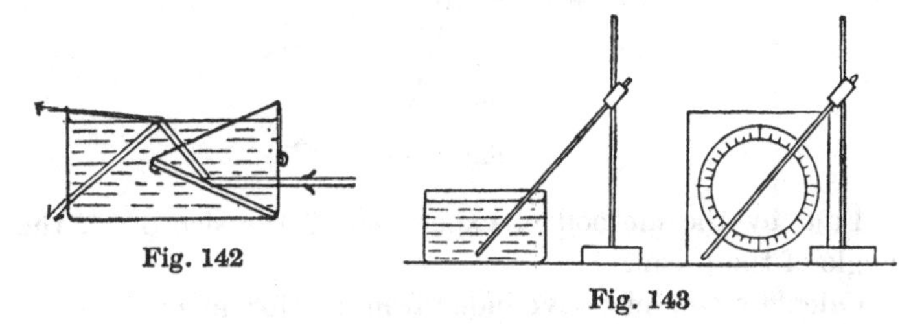

Fig. 142

Fig. 143

The angle made by the incident beam and the normal is measured by first adjusting in the liquid a knitting-pin held in a stand, and then placing the pin in front of a vertical protractor (see Fig. 143).

EXPERIMENT 9

The deviation produced by a prism.

Obtain a beam with one edge on a line drawn on paper. Place the apex of a prism in the path of the beam, draw round the prism and mark in the refracted beam.

Now push the drawing-board across so that a new position of the incident beam is obtained parallel with the former one. Note that the refracted beam retains a position parallel to its former one.

Remove the prism, produce the lines until they meet and measure the angle of deviation.

EXPERIMENT 10

Minimum deviation of prism and μ.

Obtain a deviated beam by means of the apex of a prism so that part of the original beam is still visible beyond the prism (see Fig. 144).

Rotate the prism until the angle of deviation has its minimum value, then mark in the lines and measure D, the angle of Minimum Deviation.

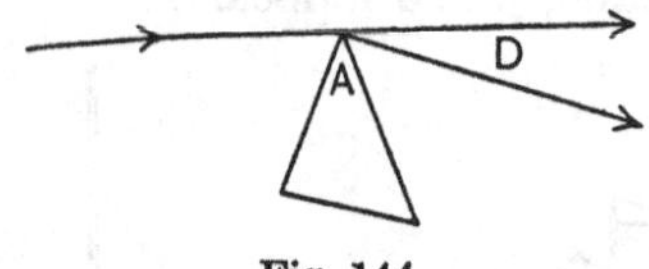

Fig. 144

Find by the method of experiment 6 the size of A, the angle of the prism.

Calculate the refractive index from the formula

$$\mu = \frac{\sin \frac{A+D}{2}}{\sin \frac{A}{2}}.$$

Repeat for the other angles of the prism.

EXPERIMENT 11

Critical angle, using an air film.

Construct an air film by cementing together two microscope slides with a ring of lead paper between them at the

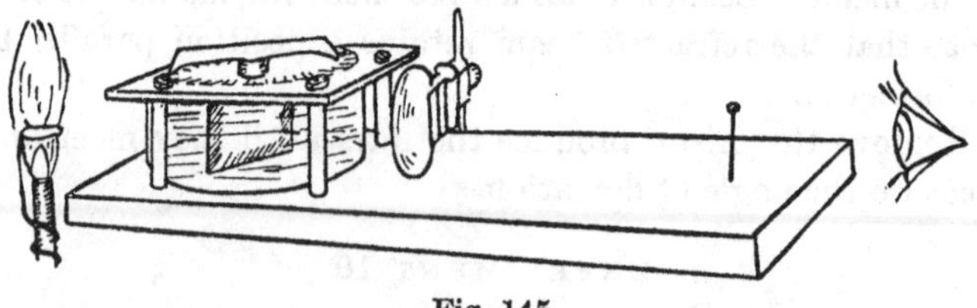

Fig. 145

edges. Mount it vertically in a glass crystallising dish as shown, with a horizontal pointer attached, the ends of which pass over a graduated circle. A convex lens is placed in front of the dish, and a vertical pin is put at the principal focus of the lens. A sodium flame is then placed on the far side of the dish. When the air film is slowly rotated, a point is reached at which the light is cut off, and to an observer the edge of the darkness is seen to pass across the lens as a straight line. When it reaches the pin the reading of the pointer on the divided circle is taken. The film is then rotated in the opposite direction until once again the light is cut off, and the reading is taken as before. The angle between the two positions is twice the critical angle.

Note. Light is totally reflected on tending to pass from water to air when an angle slightly greater than the critical angle is reached. The thin parallel-sided glass plate makes no difference, its sole effect being to give transmitted light a negligible lateral shift.

EXPERIMENT 12

Concave and convex mirrors.

For this and the following experiment a piece of stiff card about 8 in. long and about an inch wider than the lenses or mirrors is needed. A slot just big enough to take the lens or mirror is cut transversely about 3 in. from one end. The card is then pinned on to a small wood block, the lens or mirror slipped half-way through the slot as shown, and held in position by rubber bands.

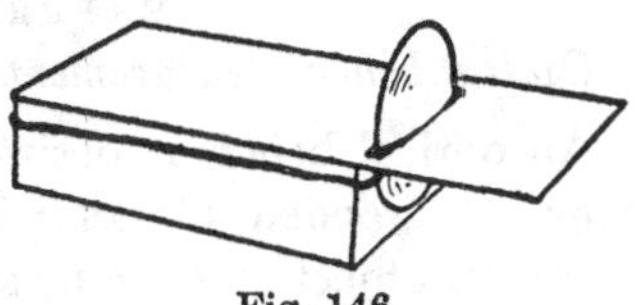

Fig. 146

Using first a concave mirror facing towards the long half of the card, show the following:

(*a*) Any ray originally parallel to the principal axis of the mirror is reflected through the principal focus.

(*b*) An incident ray through the centre of curvature comes back along the same path.

(*c*) An incident ray to the pole of the mirror, and its reflected ray, make equal angles with the axis.

(*d*) An incident ray through the focus is reflected parallel to the axis.

Repeat, using a convex mirror.

EXPERIMENT 13

Concave and convex lenses.

Substituting a convex lens for the mirror, and letting the short side of the card face the incident light, show the following:

(*a*) Rays parallel to the axis are refracted through the principal focus.

(*b*) Rays through the centre of the lens are unchanged in direction.

(*c*) Rays through the focus are refracted parallel to the axis.

Repeat, using a concave lens.

EXPERIMENT 14

Optical bench measurements.

An optical bench is desirable, but a metre rule held on edge by wooden blocks will serve. A very satisfactory luminous object is made by soldering a small piece of copper gauze on to a piece of brass in which is cut a circular aperture 1 cm. in diameter, and clipping this on to the condenser tin.

To find the focal length of a concave mirror.

Clip a piece of white paper with a hole in it on to the front of the brass carrying the gauze. Place the "object" at some convenient point against the scale, and hold the mirror in a stand opposite to it. Adjust the mirror until a sharp image is formed on the paper beside the object. Then the distance between gauze and mirror, which can be read off on the scale, is the radius of curvature, and equals $2f$.

EXPERIMENT 15

To find the focal length of a convex lens.

(*a*) Reflection method.

Proceed exactly as in the preceding experiment, but instead of the mirror use the given lens with a strip of plane mirror placed behind it.

The distance between the lens and the image formed beside the object is the focal length of the lens.

EXPERIMENT 16

To find the focal length of a convex lens.

(*b*) Conjugate foci.

Place the gauze and a screen at some convenient distance apart, say 80 cm. Hold the lens in a stand near the gauze and move it away until a sharp image is formed on the screen. The gauze and image then occupy positions known as conjugate foci and are interchangeable.

Move the lens forward until another sharp image is formed. Measure (*a*) the distance between screen and gauze, and (*b*) the distance between the two positions of the lens.

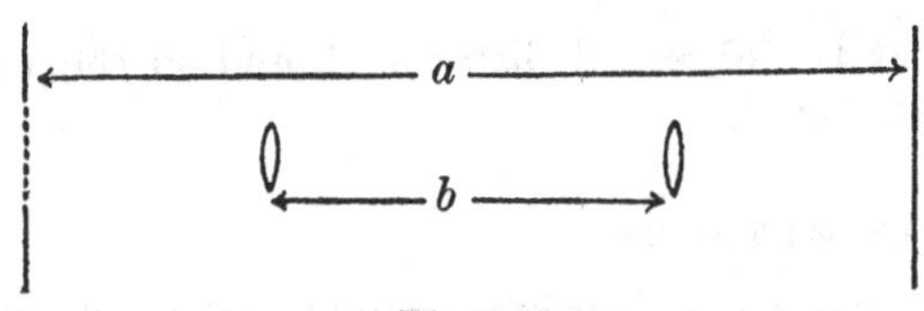

Fig. 147

From the diagram

$$a = v + u,$$
$$b = v - u.$$

Solve these equations and find f from the formula

$$\frac{1}{f}=\frac{1}{v}-\frac{1}{u}.$$

N.B. Using the sign convention on p. 108, u is positive and v negative.

EXPERIMENT 17

To find the focal length of a convex lens.

(*c*) Magnification method.

Obtain two scales on glass, preferably reproduced photographically from the same negative. Clip one in front of the condenser to act as a luminous object, and by means of the lens obtain a real image on the other scale as screen. By comparing where the lines coincide, measure the magnification, m_1.

Keeping the lens fixed, move the screen back some convenient distance, d, and adjust the object until a sharp image is again obtained. Find the new magnification, m_2.

Now $$m_1=\frac{v_1}{u_1}=\frac{f-v_1}{f}$$

and $$m_2=\frac{f-v_2}{f}.$$

So that $$m_1-m_2=\frac{v_2-v_1}{f}.$$

v_2-v_1 is the observed distance d, and so the value of f can be calculated.

PARALLAX METHODS

For parallax methods it is essential to have a clearly visible object. This is obtained by fixing a coil of stout copper wire around a bunsen with an end vertical in the flame. A red-hot, vertical line is thus obtained.

Results are incomparably better if black glass mirrors are used. These are expensive, but equally efficient ones can be made by cleaning off the silvering from old and damaged mirrors and painting the backs and edges with black paint. Such mirrors reflect from the front surface, thus overcoming one great objection to ordinary silvered mirrors, and also do not show up a lot of unwanted detail of surrounding objects.

EXPERIMENT 18

Position of image behind a plane mirror.

Arrange the luminous wire in front of the strip of black mirror at some convenient distance. On looking into the mirror the virtual image can be seen.

Place a mounted knitting-pin behind the mirror and adjust its position until it is found, by parallax, to coincide with the part of the image that is visible.

Note that the line between wire and pin is at right angles to the mirror and that they are equidistant from its front surface.

EXPERIMENT 19

Focal lengths of spherical mirrors.

(*a*) Concave.

Use a black concave mirror. Set up the luminous wire in front of it and adjust the position of a mounted knitting-pin until it coincides with the inverted real image.

Measure distance of object and image from mirror and substitute in the equation

$$\frac{1}{v}+\frac{1}{u}=\frac{1}{f}.$$

As before u must have a positive sign. Moreover the image is on the same side of the mirror as the object, and so v is positive also.

(*b*) Convex.

Proceed as above. The image is virtual, erect and behind the mirror. For this latter reason v must be allotted a negative sign when substituting in the equation.

EXPERIMENT 20

Focal lengths of lenses.

(*a*) Convex.

(i) Set up the lens with the luminous wire to one side. Look from the other side for the real inverted image and adjust a pin to coincide with it. Measure v and u. Allot a positive sign to u and a negative sign to v, and solve the equation

$$\frac{1}{v}-\frac{1}{u}=\frac{1}{f}.$$

(ii) Use method of conjugate foci (see p. 171).

(*b*) Concave.

Select a convex lens that is more powerful than the given concave lens, and measure its focal length f_1. Clip the two lenses together. They will act as a less powerful convex lens and their combined focal length, F, can be measured by any of the foregoing methods.

Then the focal length f_2 of the concave lens can be found from

$$F=f_1+f_2.$$

EXPERIMENT 21

Production of a pure spectrum.

(*a*) Place a slit in front of the condenser and by means of a lens obtain a sharp image of it on a screen. Interpose the prism, and adjust it into the position of minimum deviation. Then a pure spectrum will appear on the screen.

(b) Illuminate the slit, place the prism in position and obtain a blurred patch of refracted light on a screen. Place a convex lens between the prism and the screen and adjust it until the pure spectrum is obtained.

(c) Proceed as in (b), but instead of a screen place the eye in the path of the refracted light. A virtual spectrum is seen, and this can be made very pure by making the slit very narrow.

EXPERIMENT 22

The magic lantern.

Use the lamp and condenser, and clip a brass plate with a circular aperture in front. By means of a 10-cm. focal-length lens obtain a magnified image on a screen.

Now hold a small object such as a microscope slide against the aperture. A magnified inverted image will be obtained.

EXPERIMENT 23

The real and apparent thickness and μ.

By means of a travelling microscope and a glass block the value of μ may be found with a fair degree of accuracy. Mark a cross on paper and focus the microscope on it. Take the reading. Stand the glass block on the paper and raise the microscope until the cross is once again in focus. Again take the reading.

Finally shake some chalk dust on to the top of the glass and raise the microscope until this is in focus, when the reading is again taken.

The difference between the first and third readings gives the real thickness of the block, and that between the second and third gives the apparent thickness.

Then
$$\mu = \frac{\text{real thickness}}{\text{apparent thickness}}.$$

TABLE THREE

Velocity constants of various media for yellow light

Medium	Refractive index	Velocity constant
Crown glass	1·5170	0·6597
Rock-salt	1·5443	0·6476
Quartz	1·5443	0·6476
Flint glass	1·6499	0·6062
Diamond	2·4170	0·4137
Water	1·3330	0·7503
Glycerine	1·4700	0·6803
Carbon disulphide	1·6320	0·6128

EXAMPLES

PLANE REFLECTION AND REFRACTION

1. Draw a line 3 in. long to represent a plane mirror. Two inches from the line mark a spot to represent a point object and draw any four lines to represent incident rays falling on the mirror. Using your knowledge of the laws of refraction, find the reflected rays and produce them until they meet. What does the point of intersection represent? How far is it from the original point?

2. Two plane mirrors are inclined at 60°. A large printed R is placed between them near to one reflecting surface. Construct a diagram showing the position and appearance of all the images that will be formed.

3. A candle is placed in front of two mirrors at right angles. Make a diagram showing all the images produced and draw the rays showing how light forming the second order image reaches the eye.

4. What is the shortest length for a plane mirror in which a man 5 ft. 9 in. in height can see a complete image of himself?

5. Construct a diagram to scale, showing by means of a ray the path of light through a parallel-sided glass block whose thickness is 5 cm. and whose refractive index is 1·5. The angle of incidence is 40°. Measure the lateral displacement.

6. Find an expression showing the relation between the actual and apparent thickness of a block of glass. If a mark on paper appears to be raised $\frac{3}{4}$ in. when viewed from above through a cube of glass whose refractive index is 1·6, find the actual thickness of the glass.

7. On a very hot sunny day it is a common experience to see what appear to be pools of water on the crests of shallow hills on a smooth road. Actually there is no trace of water present. Explain this, using a diagram.

8. What is the critical angle of a medium, and how is it related to the refractive index? What will be the critical angle of water? ($\mu = 1{\cdot}33$.)

9. How can the production of a luminous cascade (see Fig. 39, p. 45) be reconciled with the fact that light always travels in straight lines? Draw a diagram showing the actual path of light in the jet of water.

PRISMS

10. What is the angle of minimum deviation given to a narrow beam of yellow light by a crown glass prism whose angle is 60°? (μ for crown glass = 1·5.)

11. A prism whose angle is 45° gives a minimum deviation of 27° 44′ to a beam of yellow light. What is the refractive index of the material of the prism?

12. It is required to make a prism of a flint glass ($\mu = 1{\cdot}75$) which shall impart the same minimum deviation to yellow light as does a crown glass prism ($\mu = 1{\cdot}5$) with an angle of 40°. What must be the angle of the second prism?

N.B. $\text{Sin}\,(x + y) = \sin x \cos y + \cos x \sin y$.

13. A prism is to be made of a glass with a refractive index of 1·64 to give a deviation of 30°. What must be the angle of the prism?

N.B. See note to question 12.

14. A narrow beam of light falls on the apex of a prism, and the angle between the two reflected beams is 60°. The angle of minimum deviation for the apex using yellow light is 18° 55′. What is the refractive index of the material?

15. What is the total deviation imparted to yellow light by a train of four prisms (such as is shown in Fig. 89) so placed that the light passing through each suffers minimum deviation, if the angle of each prism is 50° and the refractive index of the glass is 1·57?

SPHERICAL MIRRORS

16. An object 2 cm. high is placed in front of a concave mirror. When the object is 50 cm. away from the mirror the image is 4 cm. long. What is the focal length of the mirror?

17. A pin is placed in front of a convex mirror and a convex lens is put between them. For a certain position of the lens a real image is obtained coinciding with the pin. Draw a diagram showing the rays that have brought this about.

18. In the previous example, if the focal length of the lens is 20 cm., the distance from lens to mirror 5 cm., and from lens to pin 40 cm., what is the radius of curvature of the mirror?

19. An illuminated gauze is placed in front of a concave mirror and adjusted until an image is obtained coinciding with the object. The mirror is moved 10 cm. nearer to the object, and the screen has then to be moved 30 cm. further away before a sharp image is obtained. What is the focal length of the mirror?

20. Four mirrors were tested on the optical bench and the following results were obtained:

(*a*) $v =$ 75 cm. $u =$ 50 cm.
(*b*) $v =$ 125 cm. $u =$ 70 cm.
(*c*) $v =$ 37·5 cm. $u =$ 75 cm.
(*d*) $v =$ 64·6 cm. $u =$ 105 cm.

In each case find the value of f.

21. An object 5 cm. high is placed 12 cm. in front of a spherical mirror and an image is formed 60 cm. behind the mirror. What is the nature of the mirror and what its focal length? What will be the size of the image?

22. Using the mirror of the preceding question, if an image $1\frac{3}{4}$ cm. long is required where must the object be placed and where will the image be formed?

LENSES

23. Show by means of diagrams how you could use a convex lens to obtain (*a*) a magnified inverted image and (*b*) a magnified erect image of a small object such as a pin. In what manner will these images differ?

24. A convex lens is found to give a real inverted image on a screen placed 35 cm. from it of a luminous object 44 cm. from the lens. What is the focal length of the lens?

25. A lens is required to give an image of a small object magnified twelve times, at a distance of 4 ft. 6 in. away from the lens. What must be the focal length of the lens?

26. For a given convex lens the following readings were obtained on an optical bench:

v	− 17·15	− 18·27	− 20	− 23·1	− 30	− 60	+ 60	+ 8·57	
u	40	35	30	25	20	15	10	5	cm.

Draw a graph of v against u, and find the value of f.

N.B. Find when image and object are equidistant from the lens.

27. Four convex lenses were placed in turn on an optical bench. The object used was a luminous gauze which was kept at a distance of 40 cm. from the lens holder. The screen gave a sharp image at the following distances respectively: (*a*) 30 cm., (*b*) 45 cm., (*c*) 92 cm., (*d*) 77 cm. In each case find the value of f.

28. A plane mirror is placed behind a convex lens, and the image of a pin is found to coincide with the pin itself at a distance of 21 cm. from the lens. A second lens is clipped on to the first one and the image and object now coincide at 12 cm. Where would they coincide if the second lens were used alone?

29. Find by construction and verify by calculation the position and size of the image formed by a convex lens of 12 cm. focal length, when an object 3 cm. long is placed 25 cm. from the lens.

30. What is meant by the Focal Power of a lens? Explain what is meant by saying a lens has a power of $2\frac{1}{2}$. At what distance must an object be placed from such a lens to produce an image equal in size to itself?

31. A piece of pencil, 4 cm. long and 1 cm. in diameter, is placed exactly along the axis of a convex lens of focal length 10 cm., with the nearer end 20 cm. from the lens. Calculate (*a*) the position and length of the image, (*b*) the diameter of each end of the image.

32. A lens has a focal power of 4. An object is placed 30 cm. from the lens. Find (*a*) the curvature of light from the object on reaching the lens, (*b*) the curvature of the light that has just passed through the lens, (*c*) the position of the image.

33. Find by construction the position and size of the image formed by a concave lens of focal length 20 cm. when an object 3 cm. long is placed 36 cm. from the lens. Draw the object on one side only of the optical axis. Verify your result by calculation.

34. A plano-convex lens of crown glass ($\mu = 1{\cdot}5$) has a focal length of 30 cm. What is the radius of curvature of the convex surface?

35. A bi-concave lens has a focal length of 25 cm. Each surface has a radius of curvature of 40 cm. What is the value of μ for the glass?

36. A meniscus lens has one radius of curvature equal to twice the other. The material is crown glass ($\mu = 1{\cdot}5$), and the focal length 80 cm. Find the value of each radius.

37. Find the focal length of a convex lens whose radii of curvature are 16 cm. and 21 cm. The lens is made of a glass with a refractive index of 1·6.

THE EYE AND OPTICAL INSTRUMENTS

38. What are the particular advantages possessed by those animals that have two eyes in front of their head?

39. Compare the optical system of the eye with that of the photographic camera.

How is each adapted for obtaining sharp images of objects at varying distances?

40. A man with defective sight is incapable of seeing anything clearly that is more than 36 in. away from him. What is the most probable cause of the defect and how can it be cured?

41. What is meant by "long sight"? The nearest distance at which a sufferer can see clearly is 12 ft. What lenses must he wear for reading?

42. What is meant by persistence of vision? Give an account of any well-known phenomenon depending upon this.

43. Two convex lenses, one of focal length 40 cm. and the other of focal length 6·67 cm. are combined to form a telescope. How should they be placed and what magnification would be produced? Name any objection to the use of such an instrument for viewing distant places.

44. A Galilean telescope is made of two lenses, one of focal length 50 cm. and one 10 cm. Draw a diagram showing the path of a pencil of rays through the instrument and find its magnification.

45. What is the greatest size of the image of an object an inch long when viewed through a lens of 6 in. focal length?

46. Find an expression for the magnification due to a single convex lens.

If a magnification of 6 is desired what should be the focal length of the magnifying glass?

PHOTOMETRY AND ILLUMINATION

47. Two lamps when tested are found to give equal illumination when placed at 50 and 60 cm. respectively from the photometer. Compare the strength of the two lamps.

48. A ten C.P. lamp is used to test the strength of another lamp. When the first lamp is at 25 cm. from the photometer the other has to be placed at 45 cm. from it. What is its candle-power?

49. Two lamps balance on a photometer when the distances are 40 and 60 cm. respectively. If the first lamp is put at a distance of 50 cm. where must the second be placed to give equal illumination?

50. A lamp is balanced against a standard at a distance of 72 cm. from a photometer. A glass plate is then interposed and the lamp has then to be moved 8 cm. nearer the photometer to restore balance. What fraction of the light is absorbed by the glass?

51. What is meant by a lumen? What is the relation between the candle-power of a lamp and the lumens it emits?

52. Explain the term "Foot-candle". If 10 lumens fall on an area of 2 square feet, what is the illumination in foot-candles?

53. What is the average illumination in foot-candles on the floor of a room 10 ft. square due to a 100 candle-power lamp, fitted with a reflector that throws three-quarters of the total light emitted down to the floor?

54. It is necessary to obtain an illumination of at least 15 foot-candles on a work table 6 ft. square, by means of a lamp to be placed 4 ft. above the table. What must be the candle-power of the lamp (*a*) unshaded, (*b*) with a shade throwing one-third of the total light down on to the table?

MISCELLANEOUS QUESTIONS

(*Reprinted by kind permission of the Oxford and Cambridge Joint Examination Board from recent papers set in the Oxford and Cambridge School Certificate Examination*)

55. What information as to the propagation of light may be obtained from the study of eclipses? How would you arrange an experiment to illustrate an eclipse of the sun?

56. Light from an incandescent burner falls on a white sheet. Describe and explain the nature of the shadows produced when a tennis ball is held between the burner and the sheet.

Use your answer to show the difference between total and partial eclipses of the sun.

57. Explain two of the following observations:

(*a*) The shadow of a pen-holder cast on a screen by a lamp becomes less distinct as the lamp is brought nearer to the screen.

(*b*) A pond always appears to be shallower than it really is.

(*c*) A convex lens held close to the eye appears to magnify objects placed in front of it.

Give diagrams to illustrate your answer.

58. Assuming the laws of reflection show that the image of an object produced by a plane mirror is as far behind the mirror as the object is in front. Describe an experimental method of demonstrating this.

59. What is meant by an "image" in optics?

How would you show (*a*) theoretically, (*b*) by experiment, that the image of an object formed by a plane mirror is as far behind the mirror as the object is in front?

60. Make an accurate drawing of the path of a ray of light through a parallel-sided glass block of 6 cm. thickness, the angle of incidence of the ray being 45°. Measure the lateral shift of the ray assuming that the refractive index of the glass is 1·5.

61. State the laws of refraction of light.

When a ray of light strikes a horizontal water surface at an angle it is bent downwards, but if a stick is held obliquely partly immersed in water, the part in the water appears to be bent upwards. Explain these facts as fully as possible with explanatory diagrams.

62. Explain the formation of the images which may be seen when a candle is held in front of a thick glass mirror. Which image is the brightest and which is the faintest? Illustrate your answer by a diagram.

63. Define refractive index and critical angle. Prove the relation between them for two given media.

A ray of light falls normally on one face of a glass prism of angle 60° and refractive index 1·5. On a large scale diagram draw the subsequent course of the ray.

64. Explain what is meant by the total reflexion of light. Describe how you would demonstrate this phenomenon and explain some useful application of it.

65. Describe an experiment to illustrate total reflexion. Draw a diagram to show the direction in which objects outside would appear to an eye immersed in water.

66. What do you understand by the term *critical angle*? How is the critical angle related to the refractive index of a medium? Describe some experiments to illustrate total internal reflexion.

67. A triangular glass prism (refractive index 1·5) has three polished faces, the angles between the faces being 90°, 45° and 45°. A beam of light is directed normally on one of the faces enclosing the right angle. Trace the path of the ray through the prism, giving reasons for your construction. For what purposes are such prisms employed?

68. Explain why a ray cannot always be refracted from one medium to another. Describe and explain two commonly observed effects of the refraction of light.

69. State the laws of refraction of light.

A small luminous object is placed at a depth of 10 cm. below the surface of a sheet of water. Draw a full-size diagram to show the paths of rays from the object which meet the surface of the water at angles of 90°, 60° and 20° respectively. The refractive index of water is 4/3.

70. Explain, with a diagram, what a fish at the bottom of a stream might see by looking upwards.

71. Explain why a lake of clear water appears to be shallower than it actually is.

If a certain part of a lake is 6 ft. deep, what is its apparent depth when viewed vertically? (μ for water = 4/3.)

72. Of what practical use is a knowledge of the velocity constants of substances? Explain how you would determine the velocity constant for white light for a piece of glass.

73. Describe an accurate method of determining the refractive index of a liquid.

74. Explain what may happen to sunlight that falls on the face of a triangular glass prism.

Underground rooms are often illuminated by means of glass prisms fixed in the pavement outside the house. Explain how the device works.

75. Describe an arrangement for producing a pure spectrum on a screen.

Describe and account for the differences between the spectra of (*a*) an incandescent solid, (*b*) an incandescent gas, (*c*) sunlight.

76. How would you arrange a lamp, slit, lens and prism to form a pure spectrum on a screen?

Draw a clear diagram to show the path of a pencil of white light through the system.

77. Account for any two of the following: (i) the appearance of a straight rod dipped obliquely into water; (ii) umbra and penumbra; (iii) the fact that colours may match by gaslight but clash when seen by daylight.

78. Explain clearly your reasons for believing:

(*a*) That sunlight travels in straight lines.

(*b*) That it is composed of lights of different colour.

(*c*) That the lights of different colour travel to the earth with the same velocity.

79. What is a spectrum? How is it possible to infer the presence of various elements in the sun from observation of its spectrum?

80. Describe how you would produce a pure spectrum.

Give a brief account of a method of investigating the heating effect of the various parts of the spectrum of sunlight and state the results you would expect.

81. What is meant by the *dispersion of light*?

Describe how you would demonstrate this phenomenon, and how you would show that the dispersive power of carbon bisulphide is greater than that of water.

82. Show that the focal length of a concave mirror is equal to half its radius of curvature. How would you find the focal length of such a mirror experimentally?

The reflecting mirrors of motor cars are made convex. Why is this? Draw a diagram for such a mirror, showing the positions of the eye, the object and the image seen.

83. Explain what is meant by the principal focus of a spherical mirror, and find from first principles its approximate position in the case of a concave mirror. At what distance must an object be placed in front of a concave mirror of 20 cm. radius of curvature in order to produce a real image of twice the linear dimensions of the object?

84. Define the *principal focus* of a concave mirror and, from the laws of reflexion of light, deduce its position relatively to the centre of curvature and to the pole of the mirror.

An object placed 10 cm. in front of a concave mirror gives an erect image at a distance of 20 cm. from the mirror. What is the position of the principal focus?

85. Distinguish between virtual and real images.

A concave mirror has a radius of curvature of 20 cm. Where must the object be placed in order that its image formed by the mirror shall be at a distance of 30 cm. from the mirror (*a*) when the image is real, (*b*) when the image is virtual?

86. An object 2 cm. high is placed 25 cm. in front of a concave mirror of radius of curvature 40 cm. What is the nature, position and size of the image formed? Explain why images formed by lenses sometimes have coloured edges while images formed by spherical mirrors are free from colour.

87. When a finger is held some way from the bowl of a spoon an inverted image is seen in the bowl; but when the finger is held close to it the image is erect. Explain these facts.

88. Distinguish between a real and a virtual image.

How far from a concave mirror of radius 3 ft. would you place an object to give an image magnified three times? Would the image be real or virtual?

89. How could you obtain a virtual image of an object using (*a*) a convex lens, (*b*) a concave mirror?

Illustrate your answer by diagrams.

90. Draw diagrams to show how (*a*) a real, (*b*) a virtual image may be formed by a convex lens.

An object 1 cm. in height is placed 15 cm. from a lens, and the image formed by the lens is 3 cm. in height and erect. Find the focal length of the lens.

91. Define the *focal length* of a lens.

A lens of 10 cm. focal length is used to form a real image on a screen placed 20 cm. from the lens. Draw a diagram to scale, showing the paths of rays from a point on the object through the lens and, from your diagram, find the distance of the object from the screen.

92. State the Laws of Refraction of Light.

A convex lens of 20 in. focal length forms an image of an arrow which lies *along* the axis of the lens with its mid-point 30 in. from the lens. The length of the arrow is 2 in. Find the length of the image.

93. What factors determine the focal length of a lens?

A source of light and a screen are placed 250 cm. apart. What should be the focal length of a lens, and where should it be placed in order to form on the screen a real image of the source magnified four times?

94. Under what circumstances will a convex lens produce a magnified image? Give diagrams in each case to show how the image is formed.

An object is placed 12 cm. from a convex lens of 10 cm. focal length. What is the magnification of the image?

95. It is required to throw on a screen 10 ft. square a picture of a lantern slide $3\frac{1}{4}$ in. square so that the picture fills the screen. If the distance of the screen from the lantern lens is 60 ft., what focal length of lens would be most convenient?

96. Explain the action of a convex lens when used as a simple microscope. What is meant by its *magnifying power*?

Calculate the magnifying power of a convex lens of 5 cm. focal length when held close to an eye whose least distance of distinct vision is 25 cm.

97. Draw a diagram to illustrate the construction of the human eye. What sort of spectacles are required to correct for "short sight"? Give reasons.

98. Explain the use of a convex lens as a magnifying glass.

The focal length of a magnifying glass is 2 in. It is held at a distance of half-an-inch from an eye whose distance of most distinct vision is 10 in. Find the distance at which the object should be placed.

99. Describe the defects of vision known as "long-sight" and "short-sight", and explain how they may be corrected by lenses.

A man's shortest distance of distinct vision is 3 metres. What spectacle lenses does he require to enable him to read a book at a distance of 25 cm.?

100. Explain how two lenses may be combined so as to form a simple telescope. Draw a diagram showing the path through a telescope of a beam of light from the top of a distant object, indicating the position and the character of the image seen by the eye.

101. Describe the optical arrangements of the eye and compare them with those of a photographic camera.

102. Draw a diagram to show the course of a pencil of rays through a simple telescope so as to explain the formation of the image seen by the eye. Define the *magnifying power* of a telescope and state the factors upon which this depends.

103. Describe lens combinations which may be used (*a*) for the magnification of small objects and (*b*) for the examination of distant objects. Illustrate your answer by diagrams.

104. Discuss the physical similarities and differences between radiant heat and light.

ANSWERS TO NUMERICAL EXAMPLES

4. 2 ft. $10\frac{1}{2}$ in.
6. 2 in.
8. 48° 45′.
10. 37° 10′.
11. 1·55.
12. 27° 36′.
13. 42°.
14. 1·6.
15. 132° 28′.
16. 16·67 cm.
18. 35 cm.
19. 15 cm.
20. 30; 50; 25; 40 cm.
21. 15 cm.; 25 cm.
22. 117·8 cm.; 41·23 cm.
24. 19·5 cm.
25. 4·15 in.
26. – 12 cm.
27. 17·15; 21·18; 27·87; 26·32 cm.
28. 28 cm.
29. 23·08 cm.; 2·97 cm.
30. 80 cm.
31. 2·86; 1; 0·86 cm.
32. – 3·33; 0·67; 150 cm.
33. 45 cm.; 3·75 cm.
34. 15 cm.
35. 1·8.
36. 20 cm.; 40 cm.
37. 15·135 cm.
40. Concave; 36 in.
41. Convex; 10·75 in.
43. Nearly 46·67 cm. × 6.
44. × 5.
45. 2·67 in.
46. 5 cm. or 2 in.
47. 25 : 36.
48. 32·4 C.P.
49. 75 cm.
50. 0·21.
51. 4π × C.P.
52. 5 ft.-candles.
53. 3π ft.-candles.
54. 375 C.P. 128·8 C.P.
71. 4 ft. 6 in.
83. 15 cm.
84. 6·67 cm.
85. 15 cm.; 7·75 cm.
86. Real; 100 cm.; 8 cm.
88. 4 ft. real; 2 ft. virtual.
90. 22·5 cm.
92. 8·08 cm.
93. 40 cm. 50 cm.
94. × 5.
95. 19 in.
96. × 6.
98. 1·65 in.
99. 27·27 in. Convex.

INDEX

ERRATA

Page 50, *line* 6. For "G. O. Clarke" read "W. O. Clarke."

Page 127, *line* 9 *from foot.* For "with its greater magnification" read "requiring a greater tube-length."

Printed in the United States
By Bookmasters